This volume is one of a series of short biographies derived from *The New Grove Dictionary of Music and Musicians, second edition* (London, 2001). The four volumes that inaugurate this series were chosen by John Tyrrell as outstanding examples of the biographical articles in the new edition; they are printed here with little alteration.

Laura Macy
London, 2001

THE NEW GROVE®

MOZART

Cliff Eisen and Stanley Sadie

MACMILLAN PUBLISHERS LIMITED, LONDON

PALGRAVE, NEW YORK, NY

First published in
The New Grove Dictionary of Music and Musicians®, second edition
edited by Stanley Sadie, 2001

First published in the UK 2002 by Macmillan Publishers Limited, London

This edition is distributed within the UK and Europe
by Macmillan Publishers Limited, London.

First published in North America in 2002 by Palgrave,
175 Fifth Avenue, New York, NY

Palgrave is the new global publishing imprint of St. Martin's Press LLC Scholarly and Reference Division and Palgrave Publishers Ltd. (formerly Macmillan Press Ltd.)

British Library Cataloguing in Publication Data
The New Grove Mozart (The New Grove composer biographies series)
1. Mozart, Wolfgang Amadeus, 1756–1791 2. Composers – Austria – Biography
I. Sadie, Stanley, 1930– II. Tyrrell, John
780.9'2

ISBN 0-333-80408-2

Library of Congress Cataloguing-in Publication Data
The New Grove Mozart : the New Grove composer biographies / edited by Cliff Eisen and Stanley Sadie
p. cm. - (Grove music)
Includes biographical references (p.) and index
ISBN 0-312-23325-6 (pbk.)
1. Mozart, Wolfgang Amadeus, 1756–1791. 2. Composers–Austria–Biography. I. Title: New Grove composer biographies. II. Eisen, Cliff, 1952– III. Sadie, Stanley. IV. Series.
ML410.M9 N267 2000
780'.92–dc21
[B] 00-034029

Contents

Abbreviations

General

A – alto
acc. – acompaniment, accompanied by
addl – additional
addn(s) – addition(s)
ad lib – ad libitum
Ag – Agnus Dei
anon. – anonymous
ant(s) – antiphon(s)
appx(s) – appendix(es)
arr(s) – arrangement(s), arranged by/for
attrib. – attribution(s), attributed to; ascription(s), ascribed to
Aug – August
aut. - autumn
b – born
b – bass [instrument]
B – bass [voice]
Bc – basso continuo
bn – bassoon
c – circa
CA – California
cant(s). – cantatas
cl – clarinet
conc(s). – concerto(s)
CT – Connecticut
contd – continued
Cr - Credo
d – died
DC – District of Columbia
db – double bass
Dec – December
diss. – dissertation
ed(s). – edittor(s), edited (by)
edn(s) – edition(s)
eng hn – English horn
facs. – facsimile(s)
Feb – February
fl – flute
Fr – French
frag(s). – fragment(s)
Ger. – German
Gl – Gloria
glock – glockenspiel
grad(s) – gradual(s)
hn – horn
hpd – harpsichord
Hung. – Hungarian
ibid. – ibidem [in the same place]
IL – Illinois
IN – Indiana
inc. – incomplete
incl. – includes, including
inst(s) – instrument(s)
int(s) – intermezzo(s), introit(s)
It. – Italian
Jg. – Jahrgang [year of publication/ volume]
Jan – January
kbd – keyboard
Ky – Kyrie
Lat. – Latin
MA – Massachusetts
mand – mandolin
Mar – March
movt(s) – movement(s)
MS(S) – manuscript(s)
NC – North Carolina
NJ – New Jersey
no(s). – number(s)
Nov – November
NY – New York
ob – oboe, *opera buffa*
Oct – October
off(s). – offertory, offertories
op(p). – opus, opera [plural of opus]
orat(s) – oratorio(s)
orch – orchestra(tion), orchestral
orig. – original(ly)
org – organ
os – *opera seria*
ov(s). – overture(s)
perc – percussion
perf(s). – performance(s), performed (by)
pf – piano [instrument]
pic – piccolo
pt(s) – part(s)
pubd – published
qnt(s) – quintet(s)
R – photographic reprint [edn of score or early printed source
rec – recorder
recit(s) – recitative(s)
red(s). – reduction(s), reduced for
repr. – reprinted
rev(s). – revision(s), revised by/for
S – soprano
San – Sanctus
Sept – September
Spl – Singspiel
St – Saint
str – string(s)
sum. – summer
suppl(s). – supplement(s), supplementary
sym(s). – symphony (symphonies), symphonic
T – Tenor
timp – timpani
tpt – trumpet
trans. – translation, translated by
trbn – trombone
U. – University
unattrib. – unattributed
unperf. – unperformed
va – viola
vc – cello
vn – violin
vol(s). – volume(s)
vs – vocal score
wint. – winter

Bibliographic

19CM – 19th Century Music
AcM – Acta Musicologica
AnMc, AnMc – Analecta musicologica
BMw – Beiträge zur Musikwissenschaft
CMc – Current Musicology
COJ – Cambridge Opera Journal
EM – Early Music
GSJ – Galpin Society Journal
HJbMw – Hamburger Jahrbuch für Musikwissenschaft
IMSCR – International Musicological Society: Congress Report [1930–]
JAMS – Journal of the American Musicological Society
JM – Journal of Musicology
JMR – Journal of Musicological Research
JRMA – Journal of the Royal Musical Association
MAn – Music Analysis
Mf – Die Musikforschung
MISM – Mitteilungen der Internationalen Stiftung Mozarteum
MJb – Mozart-Jahrbuch [Salzburg, 1950–]
ML – Music & Letters
MQ – Musical Quarterly
MR – Music Review
MT – Musical Times
NOHM – The New Oxford History of Music (Oxford, 1954–90)
NZM – Neue Zeitschrift für Musik
ÖMz – Österreichische Musikzeitschrift
OQ – Opera Quarterly
PRMA – Proceedings of the Royal Musical Association
RdM – Revue de musicologie
RMI – Rivista musicale italiana
SMH – Studia musicologica Academiae scientiarum
SMw – Studien zur Musikwissenschaft
ZIMG – Zeitschrift der Internationalen Musik-Gesellschaft

Library Sigla

A-GÖ – Austria, Göttweig, Benediktinerstift, Musikarchiv
A-Sm – Austria, Salzburg, Internationale Stiftung Mozarteum, Bibliotheca Mozartiana
A-Ssp – Austria, Salzburg, Erzabtei St Peter, Musikarchiv
CZ-Pu – Prague, Národní Knihovna
D-Ahk – Germay, Augsburg, Heilig-Kreuz-Kirche, Dominikanerkloster, Bibliothek
D-OF – Germany, Offenbach am Main, Verlagsarchiv André
D-WEY – Germany, Weyarn, Pfarrkirche, Bibliothek
GB-Lbl – London, British Library
H-KE – Hungary, Keszthely, Helikon Kastélymúzeum, Könyvtár
I-Baf – Italy, Bologna, Accademia Filarmonica, Archivio
I-Bc – Italy, Bologna, Civico Museo Bibliografico Musicale
RUS-SPit – Russian Federation, St Petersburg, Rossiyskiy Institut Istorii Iskusstv

List of Illustrations

The story goes that when the conductor Otto Klemperer was asked to list his favourite composers, Mozart's name was not mentioned. On being taxed with this by his interlocutor, Klemperer, it is said, replied 'Oh, I thought you meant the others'. That little anecdote illustrates how Mozart is now regarded, almost as a matter of course: as the supreme composer, rising effortlessly above all other contenders. Whether or not such a competition is valid, there is little argument about his pre-eminence, especially as a composer of operas, concertos, symphonies and chamber music. His style essentially represents a synthesis of many different elements, which coalesced in his Viennese years, from 1781 on, into an idiom now regarded as a peak of Viennese Classicism. The mature music, distinguished by its melodic beauty, its formal elegance and its richness of harmony and texture, is deeply coloured by Italian opera though also rooted in Austrian and south German instrumental traditions. Unlike Haydn, his senior by 24 years, and Beethoven, his junior by 15, he excelled in every medium current in his time.

---

1. Ancestry and Early Childhood

Johann Chrysostom Wolfgang Amadeus Mozart was born in Salzburg on 27 January 1756, and baptized the following day at St Rupert's Cathedral. His names were given as Joannes Chrysostomus Wolfgangus Theophilus. The first two names record that 27 January was the feast day of St John Chrysostom, while Wolfgangus was the name of his maternal grandfather and Theophilus a name of his godfather, the merchant Joannes Theophilus Pergmayr; Mozart sometimes preferred the Latin form, Amadeus, but more frequently Amadè, Amadé or the German form Gottlieb. He was the seventh and last child born to Leopold Mozart (*b* Augsburg, 14 Nov 1719; *d* Salzburg, 28 May 1787) and his wife Maria Anna, née Pertl (*b* St Gilgen, 25 Dec 1720; *d* Paris, 3 July 1778); only he and the fourth child, Maria Anna ('Nannerl', *b* Salzburg, 30/31 July 1751; *d* Salzburg, 29 Oct 1829), survived.

The name Mozart (spelt in a variety of forms including Mozarth, Mozhard and Mozer) is first recorded for a Heinrich Motzhart in Fischach in 1331, and appears in other villages south-west of Augsburg, notably Heimberg, from the 14th century; the paternal ancestry of the family has been traced with some certainty to Ändris Motzhart, who lived in the Augsburg area in 1486. Several early members of the family were master masons (i.e. architects), builders, craftsmen and sculptors; two, in the late 16th and early 17th centuries, were artists. Mozart's great-grandfather David (*c*1620–1685) was a master mason, his grandfather Johann Georg (1679–1736) a master bookbinder in Augsburg. His mother's family came mainly from the Salzburg region and followed middle-class occupations. Her father, Wolfgang Nikolaus Pertl, held important administrative and judicial posts at Hüllenstein, near St Gilgen, but a bout of ill-health pushed him into debt and his family was left destitute.

Leopold had married Maria Anna on 21 November 1747. A composer, violinist and theorist, he was well known even before the publication of his important *Versuch einer gründlichen Violinschule* (1756), performing his works, teaching violin, directing the court music, arranging for the purchase of music and musical instruments, and attending to numerous other details as part of his court duties. Nevertheless, the recognition of his son's genius ('the miracle which God let be born in Salzburg' as he described the boy) must have struck him with the force of a divine revelation and he felt his responsibility to be not merely a father's and teacher's but a missionary's as well. The numerous journeys that Leopold undertook, at first with his entire family but after 1769 chiefly with Wolfgang alone, were in part responsible for his lack of further advancement at court and must have physically taxed his son. Yet Wolfgang's artistic development is unimaginable without them.

Until 1773 the Mozart family rented an apartment on the third floor of the house of Johann Lorenz and Maria Theresia Hagenauer, who had a thriving grocery business with connections in several important European cities. They also acted as bankers to the Mozarts, establishing credit networks for Leopold during the tours of the 1760s. It was to the Hagenauers that most of Leopold's early letters, now the most important source of information about Mozart's travels during the 1760s, were addressed. Many of them were intended for public circulation: Leopold was keen to impress the children's triumphs on the archbishop, the Salzburg nobility and his wide circle of friends and acquaintances.

As far as is known, Leopold was entirely responsible for the education of his children, which was by no means restricted to music but also included mathematics, reading, writing, literature, languages and dancing; moral and religious training were part of the curriculum as well. (A later biographical dictionary, B. Pillwein's *Biographische Schilderungen* (Salzburg, 1821), suggests that the court singer Franz Anton Spitzeder also gave the young Mozart musical instruction, but this assertion is uncorroborated.) Mozart showed his musical gifts at an early age; Leopold noted in Wolfgang's sister's music book (the so-called Nannerl Notenbuch, begun in 1759) that Wolfgang had learnt some of the pieces – mostly anonymous minuets and other binary form movements, probably German in origin, but also including works by Wagenseil, C.P.E. Bach, J.J. Agrell and J.N. Tischer as well as Leopold Mozart himself – when he was four. According to Leopold, Wolfgang's earliest known compositions, a miniature Andante and Allegro K1*a* and 1*b*, were written in 1761, when he was five. More substantial are the binary form minuets in F major K2 and K5 and the Allegro in B♭ K3, composed between January and July 1762.

Leopold Mozart's collaboration in Wolfgang's early works up to about 1766 was probably considerable (the trio of no.48 in the 'Nannerl Notenbuch', an arrangement of the trio of the third movement of Leopold's D major serenade, also appears as Menuet II in Wolfgang's sonata K6) although he seldom drew direct attention to it. After this time he served chiefly (but not exclusively) as proofreader and editor; until the early 1770s scarcely a single autograph of Wolfgang's is without additions or alterations in his father's hand. Even later, the attributions and dates on Mozart's autographs are frequently by Leopold, who apparently preserved his son's manuscripts with painstaking orderliness. Thus the elder Mozart fulfilled a universal function as teacher, educator and private secretary to his son, and when necessary also served as valet, impresario, propagandist and travel organizer.

Mozart's first known public appearance was at Salzburg University in September 1761, when he took a dancing part in a performance of *Sigismundus Hungariae rex*, given as an end-of-term play (*Finalkomödie*) by Marian Wimmer with music by the Salzburg Kapellmeister Ernst Eberlin. In 1762 Leopold apparently took Wolfgang and Nannerl to Munich, where they played the harpsichord for Maximilian III Joseph, Elector of Bavaria (no documentation survives for this journey, which is known only from a later reminiscence of Nannerl Mozart). A tour to Vienna lasted from September to December 1762. The children appeared twice before Maria Theresa and her consort, Francis I, as well as at the homes of various ambassadors and nobles (the empress sent the children a set of court clothes, which they wore for the well-known paintings done later in Salzburg, probably by P.A. Lorenzoni). The trip was a great success: in October the imperial paymaster presented the Mozarts with a substantial honorarium and a request to prolong their stay; the French ambassador, Forent-Louis-Marie, Count of Châtelet-Lomont, invited them to Versailles; and Count Karl von Zinzendorf, later a high state official, wrote in his diary that 'the poor little fellow plays marvellously, he is a child of spirit, lively, charming; his sister's playing is masterly'.

The family returned to Salzburg on 5 January 1763. Leopold was promoted to deputy Kapellmeister on 28 February, and that evening Mozart played at court as part of Archbishop Schrattenbach's birthday celebrations; the Salzburg court chronicle records that there was 'vocal music by several virtuosos, among whom were, to everyone's astonishment, the new vice-Kapellmeister's little son, aged seven, and daughter, aged ten, performing on the harpsichord, the son likewise on the violin, as well as one could ever have hoped of him'. On 9 June the family set out on a three-and-a-half-year journey through Germany, France, the Low Countries, England and Switzerland. It was the first of five tours undertaken during the next decade.

2. TRAVELS, 1763–73

Travelling by way of Munich, Augsburg, Ludwigsburg, the summer palace of the Elector Palatine Carl Theodor at Schwetzingen, Mainz, Frankfurt, Coblenz and Aachen, the Mozart family arrived at Brussels on 4 October 1763; in each of these places the children either performed at court or gave public concerts. From there they pressed on to Paris. The children played before Louis XV on 1 January 1764, with public concerts following on 10 March and 9 April at the private theatre of M. Félix, in the rue et porte Saint-Honoré. In Paris Mme Vendôme published Mozart's two pairs of sonatas for keyboard and violin, K6–9, his first music to appear in print.

The family arrived in England on 23 April, first lodging at the White Bear Inn in Piccadilly; the next day they moved to the house of the barber John Cousins, in Cecil Court. They played twice for George III, on 27 April and 17 May 1764 (in a letter of 28 May, Leopold enthusiastically recounted to Hagenauer the friendly greeting the king gave them at a chance meeting in St James's Park), and were scheduled to appear at a benefit for the composer and cellist Carlo Graziani on 23 May; however, Wolfgang was taken ill and was unable to perform. The Mozarts mounted their own benefit on 5 June, at the Great Room in Spring Garden; later that month Mozart performed 'several fine select Pieces of his own Composition on the Harpsichord and on the Organ' at Ranelagh Gardens, during breaks in a performance of Handel's *Acis and Galatea*. Further benefit concerts followed on 21 February and 13 May 1765. At some time during their visit to London, Mozart was tested by the philosopher Daines Barrington, who in 1769 furnished a report on him to the Royal Society (published in its *Philosophical Transactions*, lx (1771), 54–64). Barrington's tests were typical of others that Mozart was set elsewhere on the Grand Tour and, later, in Vienna and Italy:

> I said to the boy, that I should be glad to hear an extemporary *Love Song*, such as his friend Manzoli might choose in an opera. The boy ... looked back with much archness, and immediately began five or six lines of a jargon recitative proper to introduce a love song. He then played a symphony which might correspond with an air composed to the single word, *Affetto*. It had a first and second part, which, together with the symphonies, was of the length that opera songs generally last: if this extemporary composition was not amazingly capital, yet it was really above mediocrity, and shewed most extraordinary readiness of invention. Finding that he was in humour, and as it were inspired, I then desired him to compose a *Song of Rage*, such as might be proper for the opera stage. The boy again looked back with much archness, and began five or six lines of a jargon recitative proper to precede a *Song of Anger*. This lasted also about the same time as the *Song of Love*; and in the middle of

> it, he had worked himself up to such a pitch, that he beat his harpsichord like a person possessed, rising sometimes in his chair. The word he pitched upon for this second extemporary composition was, *Perfido.* After this he played a difficult lesson, which he had finished a day or two before: his execution was amazing, considering that his little fingers could scarcely reach a fifth on the harpsichord. His astonishing readiness, however, did not arise merely from great practice; he had a thorough knowledge of the fundamental principles of composition, as, upon producing a treble, he immediately wrote a base under it, which, when tried, had very good effect. He was also a great master of modulation, and his transitions from one key to another were excessively natural and judicious; he practised in this manner for a considerable time with a handkerchief over the keys of the harpsichord.

The Mozarts left London on 24 July 1765, travelling by way of Canterbury (where a concert was announced, but apparently cancelled) and Lille to Ghent and Antwerp, arriving at The Hague on 10 September. There the children gave two public concerts and played before the Princess of Nassau-Weilburg, to whom Mozart later dedicated the keyboard and violin sonatas K26–31. They moved on to Amsterdam in January, returning to The Hague for the installation of Wilhelm V on 11 March – it was for this occasion that Mozart composed the *Gallimathias musicum* K32 – and in April they set out again for Paris, arriving there in early May. The Mozarts remained in Paris for two months; their patron, Baron Grimm, who had paved their way there earlier, commented on Mozart's 'prodigious progress' since early 1764.

The final stage of the homeward journey took the Mozarts to Dijon, Lyons, Lausanne, Zürich and Donaueschingen, where they played for Prince Fürstenberg on nine evenings. From Donaueschingen they pressed on to Dillingen, Augsburg and Munich, arriving back in Salzburg on 29 November. On the day of their arrival, Beda Hübner, librarian at St Peter's, wrote in his diary (in *A-Ssp*):

> I cannot forbear to remark here also that today the world-famous Herr Leopold Mozart, deputy Kapellmeister here, with his wife and two children, a boy aged ten and his little daughter of 13, have arrived to the solace and joy of the whole town ... the two children, the boy as well as the girl, both play the harpsichord, or the clavier, the girl, it is true, with more art and fluency than her little brother, but the boy with far more refinement and with more original ideas, and with the most beautiful harmonic inspirations ... There is a strong rumour that the Mozart family will again not long remain here, but will soon visit the whole of Scandinavia and the whole of Russia, and perhaps even travel to China, which would be a far greater journey and bigger undertaking still: de facto, I believe it to be certain that nobody is more celebrated in Europe than Herr Mozart with his two children.

Leopold Mozart is often portrayed as an inflexible, if consummate, tour manager, yet much of the 'Grand Tour' was not planned in advance. When he left Salzburg, Leopold was undecided whether to travel to England; nor was it his intention to visit the Low Countries (letter of 28 May 1764). There were also miscalculations. It is likely, for instance, that the Mozarts outstayed their welcome in London: by June 1765 they were reduced to giving cheap public displays at the down-market Swan and Hoop Tavern in Cornhill (see McVeigh, G 1993). On the other hand, it is not widely appreciated how difficult travel could be at this time: routes were often unsafe and almost always uncomfortable (Leopold marvelled in a letter of 25 April 1764 at his successful crossing of the English Channel, an experience that was surely unknown to his friends in Salzburg), expenses were substantial, and he was frequently mistreated, ignored or prevented by potential patrons from performing. In a letter completed on 4 November 1763 (quotations from the Mozart family correspondence are based on the translations in Anderson, A 1938, 3/1985) he wrote from Brussels:

> We have now been kept [here] for nearly three weeks. Prince Karl [Charles of Lorraine, brother of Emperor Francis I and Governor of the Austrian Netherlands] ... spends his time hunting, eating and drinking ... Meanwhile, in decency I have neither been able to leave nor to give a concert, since, as the prince himself has said, I must await his decision.

These unexpected detours – which added nearly two years to the tour – nevertheless reaped rich musical rewards: at every stage of their travels the Mozarts acquired music that was not readily available in Salzburg or met composers and performers who did not normally travel in south Germany and Austria. At Ludwigsburg they heard Pietro Nardini (on 11 July 1763 Leopold wrote to Salzburg, 'it would be impossible to hear a finer player for beauty, purity, evenness of tone and singing quality'), and in Paris they met, among others, Johann Schobert, Johann Gottfried Eckard and Leontzi Honauer, from whose sonatas, as well as sonatas by Hermann Friedrich Raupach and C.P.E. Bach, Mozart later chose movements to set as the concertos K37 and 39–41. Their stay in London brought Mozart into contact with K.F. Abel, Giovanni Manzuoli and most importantly J.C. Bach, with whom the family became intimate and whose influence on Mozart was lifelong. Years later, when Wolfgang was in Paris, Leopold upheld Bach as a model composer (letter of 13 August 1778).

> If you have not got any pupils, well then compose something more.... But let it be something short, easy and popular ... Do you imagine that you would be doing work unworthy of you? If so, you are very much mistaken. Did Bach, when he was in London, ever publish anything but similar trifles? *What is slight can still be great*, if it is written in a natural, flowing and easy style – and at the same time bears the marks of sound

> composition. Such works are more difficult to compose than all those harmonic progressions, *which the majority of the people cannot fathom*, or pieces which have pleasing melodies, but which are *difficult to perform*. Did Bach lower himself by such work? Not at all. Good composition, sound construction, il filo – these distinguish the master from the bungler – even in trifles.

It is also safe to say that on the 'Grand Tour' Mozart began to absorb his father's opinions about various national styles and how to conduct himself in public. In Paris on 1 February 1764, Leopold wrote of the Royal Chapel at Versailles:

> I heard good and bad music there. Everything sung by individual voices and supposed to resemble an aria was empty, frozen and wretched – in a word, French – but the choruses are good and even excellent ... the whole of French music is not worth a sou.

In this he anticipated by many years Mozart's comment on 5 April 1778, when he was again in Paris, that

> at Mannheim [the choruses] are weak and poor, whereas in Paris they are powerful and excellent ... What annoys me most of all in this business is that our French gentlemen have only improved their *goût* to this extent that they can now listen to good stuff as well. But to expect them to realize that their own music is bad or at least to notice the difference – Heaven preserve us!

More importantly, perhaps, Mozart also took to heart his father's negative opinions of Salzburg, repeating them almost verbatim in his letters of the late 1770s and early 80s. As early as 19 July 1763 Leopold wrote from Schwetzingen:

> The orchestra is undeniably the best in Germany. It consists altogether of people who are young and of good character, not drunkards, gamblers or dissolute fellows.

Mozart, some 15 years later, wrote to his father (letter of 9 July 1778):

> one of my chief reasons for detesting Salzburg [is the] coarse, slovenly, dissolute court musicians. Why, no honest man, of good breeding, could possibly live with them! Indeed, instead of wanting to associate with them, he would feel ashamed of them ... [The Mannheim musicians] certainly behave quite differently from ours. They have good manners, are well dressed and do not go to public houses and swill.

Mozart remained in Salzburg for nine months. During this time he wrote three vocal works: a Latin comedy, *Apollo et Hyacinthus*, for the university; the first part of the oratorio *Die Schuldigkeit des ersten und fürnehmsten Gebots*, a joint work with Michael Haydn and Anton Cajetan Adlgasser; and the *Grabmusik* K42 (to which he added a concluding chorus with introductory recitative, *c*1773). On 15 September 1767 the family set out for Vienna. Presumably Leopold had timed this visit to coincide with the festivities planned for the marriage of the 16-year-old Archduchess Josepha to King Ferdinand IV of Naples. Josepha, however, contracted smallpox and died on the day after the wedding was to have taken place, throwing the court into mourning and inducing Leopold to remove his family from Vienna, first to Brünn (Brno) and then to Olmütz (Olomouc), where both Mozart and Nannerl had mild attacks of smallpox.

Shortly after their return to Vienna in January 1768, Leopold conceived the idea of securing for Mozart an opera commission, *La finta semplice*, but intrigues at court conspired to defeat his plan (the Mozarts' side of the story is preserved in detail in the surviving correspondence). He wrote an indignant petition to the emperor in September, complaining of a conspiracy on the part of the theatre director Giuseppe d'Affligio, who apparently claimed that Wolfgang's music was ghost-written by his father, and proving Mozart's output by including a list of his compositions to that time (see Zaslaw, A 1985). Presumably as compensation for the suppression of the opera, in December Mozart directed a performance before the imperial court of a festal mass (K139), an offertory (K47*b*, lost) and a trumpet concerto (K47*c*, lost) at the dedication ceremony of the Waisenhauskirche; the *Wienerisches Diarium* reported on 10 December 1768 that Mozart performed his works 'to general applause and admiration, and conducted with the greatest accuracy; aside from this he also sang in the motets'. That same month he completed the Symphony K48. Earlier, in October, Mozart may have given a private performance of his one-act Singspiel *Bastien und Bastienne* at the home of Dr Franz Anton Mesmer, the inventor of 'magnetism therapy' (later parodied in *Così fan tutte*).

On the return journey to Salzburg, the Mozarts paused at Lambach Abbey, where father and son both presented symphonies to the library (the controversy over the attribution of the two works, Leopold Mozart's G9 and Mozart's KAnh.221, is summarized in Zaslaw, L 1989). They arrived home on 5 January and remained there for nearly a year. *La finta semplice* was performed at court on or about 1 May, and Mozart wrote the Mass K66 in October for the first Mass to be celebrated by his friend Cajetan (Father Dominicus) Hagenauer, son of the family's Salzburg landlord. Other substantial works from this time include three orchestral serenades (K63, 99, and 100), two of which were probably intended for performance as 'Finalmusik' at the university's traditional end-of-year ceremonies, possibly some shorter sacred works (K117 and 141) and several sets of dancing minuets (K65*a* and 103; K104 and 105 are by Michael Haydn, possibly arranged by Mozart). By the age of 13, then, Mozart had achieved a significant local reputation as both a composer and a performer. On 27 October he was appointed, on an honorary basis, Konzertmeister at the Salzburg court.

Less than two months later, on 13 December, Leopold and Wolfgang set out on their own for Italy. The journey followed the now usual pattern: they paused at any town where a concert could be given or where an influential nobleman might wish to hear Mozart play. Travelling by way of Innsbruck and Rovereto, they arrived at Verona on 27 December. While there, Mozart played at the Accademia Filarmonica and had his portrait painted, probably by Saverio dalla Rosa; the piece of music shown on the harpsichord, almost certainly by him, is otherwise unknown (K72*a* but see Heartz O 1995). At Mantua, on 16 January, Mozart gave a public concert typical of his public and private performances at the time: it included a symphony by him; *prima vista* and extempore performances of concertos, sonatas, fugues, variations and arias; and a small number of works contributed by other performers. The *Gazzetta di Mantova*, in a report on the concert (19 January 1770), described Mozart as 'incomparable'.

From Mantua the Mozarts travelled to Milan where Wolfgang gave several performances at the home of Count Karl Firmian, the Austrian minister plenipotentiary, including a grand academy on 12 March that may have included the newly composed arias K77, 88 and Anh.2; presumably as a result of his performances and compositions, Mozart was commissioned to write the first opera, *Mitridate, re di Ponto*, for the carnival season in December. Father and son left Milan on 15 March, bound for Lodi (where Mozart completed his first string quartet, K80), Parma, Bologna (where they met the theorist and composer Padre Martini) and Florence, where Mozart became reacquainted with the castrato Manzuoli and newly acquainted with the English composer Thomas Linley, a boy of his own age. From there they passed on to Rome, arriving on 10 April, in time for Holy Week; Mozart made a clandestine copy of Allegri's famous *Miserere* (traditionally considered the exclusive property of the papal choir) and may have composed two or three symphonies (K81, 95 and 97). After a brief stay in Naples, where Mozart gave several concerts and heard Niccolò Jommelli's *Armida* (which he described on 5 June 1770 as 'beautiful, but much too broken up and old-fashioned for the theatre'), they returned to Rome, where on 5 July Pope Clemens XIV created Mozart a Knight of the Golden Spur. Father and son set out again on 10 July, returning to Bologna and the summer home of Count Pallavicini. There Mozart may have completed the Symphony K84, as well as some sacred works and canons, and received the libretto and cast-list for his Milan opera. Before they left Bologna he was admitted to membership of the Accademia Filarmonica; the original autograph of his test piece, the antiphon K86, has annotations by Padre Martini, suggesting that he may have had help.

Work on the composition of *Mitridate, re di Ponto* began in earnest after the Mozarts' return to Milan on 18 October 1770. The libretto, by Vittorio Amadeo Cigna-Santi, after Racine, had been set by Quirino Gasparini for Turin in 1767 and Leopold in his letters described various intrigues among the singers, including the possibility of their substituting certain of Gasparini's settings for Mozart's. In fact the setting of 'Vado incontro al frato estremo' found in the earliest scores of the opera has been discovered to be by Gasparini; apparently the primo uomo, D'Ettore, was unwilling to sing Mozart's version, which is now lost (Peiretti, J 1996). There were three recitative rehearsals, two preliminary orchestral rehearsals and two full ones in the theatre, as well as a dress rehearsal; Leopold's letter of 15 December gives the useful information that the orchestra consisted of 14 first and 14 second violins, 6 violas, 2 cellos, 6 double basses, 2 flutes, 2 oboes, 2 bassoons, 4 horns, 2 trumpets and 2 keyboards. The première at the Regio Ducal Teatro was on 26 December; including the ballets (by other composers), it lasted six hours. Leopold had not been confident that the opera would be a success, but it was, running to 22 performances.

The Mozarts left Milan on 14 January 1771, stopping at Turin, Venice, Padua and Verona before returning to Salzburg on 28 March. The 15-month Italian journey had been an extraordinary success, widely reported in the international press: on 20 March 1770 the *Notizie del mondo* of Florence carried a notice of the 'magnificent academy' given at Count Firmian's, while the Hamburg *Staats- und gelehrte Zeitung* described Mozart's 'extraordinary and precocious musical talent' in a report sent from Rome on 22 May. The same newspaper's account of Wolfgang's Venice concert of 5 March 1771 (published on 27 March) neatly sums up the professional and personal accomplishments of the tour:

> Young Mozart, a famous keyboard player, 15 years old, excited the attention and admiration of all music lovers when he gave a public performance in Venice recently. An experienced musician gave him a fugue theme, which he worked out for more than an hour with such science, dexterity, harmony and proper attention to rhythm that even the greatest connoisseurs were astounded. He composed an entire opera for Milan, which was given at the last carnival. His good-natured modesty, which enhances still more his precocious knowledge, wins him the greatest praise, and this must give his worthy father, who is travelling with him, extraordinary pleasure.

Even before their return to Salzburg in March 1771, Leopold had laid plans for two further trips to Italy: when the Mozarts were in Verona, Wolfgang was commissioned to write a serenata or *festa teatrale*, *Ascanio in Alba*, for the wedding in Milan the following October of Archduke Ferdinand and Princess Maria Beatrice Ricciarda of Modena; that same month the Regio Ducal Teatro at Milan had issued him with a contract for the first carnival opera of 1773, *Lucio Silla* (an oratorio commissioned for Padua, *La Betulia liberata*, seems never to have been performed). Accordingly, Mozart spent barely five months at home in 1771, during which time he wrote the Paduan oratorio, the *Regina coeli* K108, the litany K109 and the Symphony K110. Father and son set out again on 13 August, arriving at Milan on 21 August. They received Giuseppe Parini's libretto for *Ascanio in Alba* on 29 August; the serenata went into rehearsal on 27 September and the première took place on 17 October. Hasse's Metastasian opera *Ruggiero*, also composed for the wedding festivities, received its first performance the day before; according to Leopold, *Ascanio* 'struck down Hasse's opera' (letter of 19 October 1771), a judgment confirmed by a report in the Florentine *Notizie del mondo* on 26 October: 'The opera has not met with success, and was not performed except for a single ballet. The serenata, however, has met with great applause, both for the text and for the music'. The Mozarts remained in Milan until 5 December; Wolfgang wrote the curiously titled 'Concerto ò sia Divertimento' K113 (later revised for Salzburg performance; see Blazin, L 1992) and the Symphony K112. He also may have sought employment at court, but his application was effectively rejected by Ferdinand's mother, Empress Maria Theresa, who in a letter (12 December 1771) advised the archduke against burdening himself with 'useless people' who go 'about the world like beggars'.

The third and last Italian journey began on 24 October 1772; probably Mozart had been sent the libretto and cast-list for the new Milan opera, *Lucio Silla*, during the summer, and had also set the recitatives. On his arrival at Milan, these were adjusted to accommodate changes made by the poet, Giovanni de Gamerra. He then wrote the choruses, and composed the arias for the singers in turn, having first heard each of them so that he could suit the music to their voices. The première, on 26 December, was a mixed success, chiefly because of a patchy cast; nevertheless, the opera ran for 26 performances. In January Mozart wrote the solo motet *Exsultate, jubilate* for the primo uomo in the opera, Venanzio Rauzzini (in Salzburg, about 1780 he revised the motet, probably for the castrato soprano Francesco Ceccarelli to sing at the Dreifaltigkeitskirche; see Münster, I 1993).

Leopold and Wolfgang arrived back in Salzburg on 13 March 1773. Mozart's days as a child prodigy were over: although he later travelled to Vienna, Munich and, more importantly, Mannheim and Paris, the 1770s can fairly be described as dominated by his tenure at Salzburg. For the most part, his career as both performer and composer was focussed on his court activities and a small circle of friends and patrons in his native town.

3. SALZBURG, 1773–80

Archbishop Schrattenbach, who died on 16 December 1771, the day after Wolfgang's return from the second Italian tour, was succeeded in March

1772 by Hieronymus Colloredo. An unpopular choice whose election was bitterly contested, Colloredo sought to modernize the archdiocese on the Viennese model, but his reforms, while generally favouring cultural life in the city by attracting numerous prominent writers and scientists, met with local resistance. The court music in particular suffered, and many traditional opportunities for music-making were eliminated: the university theatre, where school dramas (the nearest Salzburg equivalent to opera) had been performed regularly since the 17th century, was closed in 1778; the Mass was generally shortened; restrictions were placed on the performance of purely instrumental music as well as some instrumentally accompanied sacred vocal music at the cathedral and other churches; and numerous local traditions, including the firing of cannons and the carrying of pictures and statues during church processions as well as the famous pilgrimage to Pinzgau, were abolished. Concerts at court were curtailed; in a letter of 17 September 1778 Leopold Mozart complained,

> Yesterday I was for the first time [this season] the director of the great concert at court. At present the music ends at around 8.15. Yesterday it began around 7.00 and, as I left, 8.15 struck – thus an hour and a quarter. Generally only four pieces are done: a symphony, an aria, a symphony or concerto, then an aria, and with this, Addio!

While these changes profoundly affected traditional composition and performance in Salzburg, they also encouraged other kinds of musical activity. In 1775 Colloredo ordered that the Ballhaus in the Hannibalgarten (now the Makartplatz) be rebuilt, at the city's expense, as a theatre for both spoken drama and opera. The first troupe to play there, directed by Carl Wahr, included in its repertory the comedy *Der Zerstreute* (after J.F. Regnard), with incidental music by Joseph Haydn (Symphony no.60, 'Il distratto'), while Gebler's tragedy *Thamos, König in Ägypten* may have been performed with incidental music by Mozart. Schikaneder's troupe visited in 1780; Mozart composed the aria KAnh.11*a* (of which only a fragment survives) for his production of *Die zwei schlaftlosen Nächte* (Edge, K 1996). Private orchestras were also established, the first of them by Colloredo's nephew, Count Johann Rudolf Czernin. Nevertheless, Colloredo's reforms served ultimately to impoverish Salzburg's musical life, and his policy of promoting Italians at the expense of local German talent – Domenico Fischietti was appointed Kapellmeister in 1772, and Giacomo Rust in 1777 – was a frequent cause for complaint. This may have been a sticking-point for Leopold Mozart in particular, who as deputy Kapellmeister since 1763 had reasonable expectations for promotion; as early as 1763 he had lamented the power and influence of Italian musicians in Germany, attributing his failure to secure an audience with Duke Carl Eugen of Württemberg to the intrigues of his Ober-Kapellmeister, Jommelli. In Paris in

1764 he wrote to Hagenauer: 'If I had *one single wish* that I could see fulfilled in the course of time, it would be to see Salzburg become a court which made a tremendous sensation in Germany with its own local people'.

Mozart composed prolifically during the early years of Colloredo's rule: between 1772 and 1774 he wrote the masses K167, 192 and 194, the litanies K125 and 195, the *Regina coeli* K127, more than a dozen symphonies (from K124 to K202), the keyboard concerto K175 (possibly for organ) and the Concertone for two solo violins K190, the serenade K203, the divertimentos K131, 166 and 205 and the Quintet K174 (presumably modelled on similar works by Michael Haydn; see Seiffert, in Eisen and Seiffert, N 1994). Financially the family prospered: in late 1773 they moved from their apartment in the Getreidegasse, where they had lodged with the Hagenauers, to a larger one, the so-called Tanzmeisterhaus, in the Hannibalplatz (now the Makartplatz). No doubt this move reflected Leopold's consciousness of their status in Salzburg society: the family was socially active, taking part in shooting parties and in constant music-making and often receiving visitors. Nevertheless, encouraged by rumours of a possible opening at the imperial court, Leopold took Wolfgang to Vienna in July 1773. Nothing came of this, but the sojourn, which lasted four months, was a productive one for Mozart: he composed a serenade (K185, possibly intended as a Salzburg Finalmusik) and six string quartets (K168–73). The more intense style of the quartets (two of which, K168 and 173, include fugal finales) has traditionally been attributed to Mozart's presumed contact with Joseph Haydn's latest quartets, in particular opp.9, 17 and 20, although it is more likely that they reflect common elements of the Viennese quartet at the time (Brown, H 1992).

Mozart returned from Vienna in late September, and with the exception of three months spent in Munich between December 1774 and March 1775 for the composition and première of *La finta giardiniera*, the libretto of which is generally thought to have been prepared by Coltellini after Goldoni, he remained in his native city until September 1777. In the absence of any sustained family correspondence, his activities can only be surmised. No doubt they included performing at court and in the cathedral, frequent musical gatherings at home, considerable social activity and composition. Among the few documented events of these years are the composition of *Il re pastore* for the visit to Salzburg of Archduke Maximilian Franz on 23 April 1774 and Mozart's participation in celebrations marking the 100th anniversary of the pilgrimage church at Maria Plain in 1774.

It was about this time that Mozart began to withdraw from the Salzburg court music, although the root cause of his dissatisfaction remains unclear. The family letters document Leopold's frustrating inability to find suitable positions for both of them; they frequently complain of longstanding troubles with

Colloredo, who is described as rude and insensitive. And there was the irritation of being outdone in the court music by Italians, who were better paid than local musicians. Yet there is no compelling evidence of Colloredo's mistreatment of the Mozarts early in his rule. Wolfgang's serenata *Il sogno di Scipione*, originally composed for the 50th anniversary of Schrattenbach's ordination, was reworked early in 1772 and performed as part of the festivities surrounding Colloredo's enthronement; on 21 August 1772 he was formally taken into the paid employment of the court, as Konzertmeister (a post he had held in an honorary capacity for nearly three years) with an annual salary of 150 gulden, while Leopold continued to run the court music on a periodic basis and was entrusted with securing musicians, music and instruments; and the Mozarts travelled to Italy, Vienna and Munich. Their discontent with Salzburg – and Colloredo's eventual rejection of them – must therefore have had grounds beyond the conditions of their employment, Colloredo's difficult personality, his attempts to reform music-making in Salzburg or his general belt-tightening.

No doubt Colloredo was displeased by Leopold's excessive pride and his superior manner (in November 1766 Leopold had written, 'after great honours, insolence is absolutely not to be stomached') and in particular by his continuing attempts to leave the court. Both in Italy (1770–71) and in Vienna (1773) Leopold had attempted to find jobs that would permit the family to quit Salzburg, and not for the first time. As early as 30 October 1762, when he was in Vienna, he wrote a thinly veiled threat to Hagenauer: 'If only I knew what the future will finally bring. For one thing is certain: I am now in circumstances which allow me to earn my living in Vienna'; and in London he was offered a post that, after much consideration, he rejected. Leopold frequently wrote of his plans in his letters home, often in cypher, to prevent them from being read and understood by the Salzburg censors. But it is likely that they were well known to Colloredo, who had good connections both in Vienna and in Italy. Maria Theresa's description of the family as like 'beggars' may have represented a common view among some of the European nobility.

Mozart's rejection of court musical life was transparent. He continued to compose church music, the primary duty of all Salzburg composers, but with little enthusiasm: his output between 1775 and 1777, including the masses K220, 257–9, 262 and 275, the litany K243 and the offertory K277, was meagre compared with Michael Haydn's. Instead, Mozart established himself as the chief composer in Salzburg of instrumental and secular vocal music. Four violin concertos (K211, 216, 218 and 219; K207 was composed earlier, in 1773) and four keyboard concertos (K238, K242 for three keyboards, K246 for two and K271, presumably for the otherwise unknown French pianist Mlle Jeunehomme), the serenades K204 and 250, the 'Serenata notturna' K239 and numerous divertimentos (including K188, 240, 247 and 252) all date from this time; he also composed several arias, including *Si mostra la sorte* K209, *Con ossequio, con rispetto* K210, *Voi avete un cor fedele* K217 and *Ombra felice ... Io ti lascio* K255. It is likely that Mozart's cultivation of instrumental music, which in many cases he wrote for private patrons rather than the court, was encouraged by Leopold, who during his heyday had been the most prominent and successful local composer of symphonies and serenades. Yet this may also have been a miscalculation. Leopold apparently failed to recognize that the conditions of musical life in the archdiocese, to say nothing of musical taste, had changed since the 1750s.

Matters came to a head in the summer of 1777. In August Mozart wrote a petition asking the archbishop for his release from employment, and Colloredo responded by dismissing both father and son. Leopold, however, felt he could not afford to leave Salzburg, and so Mozart set out with his mother on 23 September. The purpose of the journey was clear: Mozart was to secure well-paid employment (preferably at Mannheim, which Leopold had described in a letter of 13 November 1777 as 'that famous court, whose rays, like those of the sun, illuminate the whole of Germany') so that the family could move. Mozart first called at Munich, where he offered his services to the elector but met with a polite refusal. In Augsburg he gave a concert including several of his recent works and became acquainted with the keyboard instrument maker J.A. Stein; in a letter of 17 October he described Stein's pianos as damping

> ever so much better than [Späth's] instruments. When I strike hard, I can keep my finger on the note or raise it, but the sound ceases the moment I have produced it. In whatever way I touch the keys, the tone is always even. It never jars, it is never stronger or weaker or entirely absent; in a word, it is always even.

He also embarked on a relationship with his cousin, Maria Anna Thekla (the 'Bäsle'), with whom he later engaged in a scatological correspondence. Although obscene humour was typical of Salzburg (Mozart's parents sometimes wrote to each other in a similar vein), Solomon (F 1995) has argued that the relationship between Wolfgang and the Bäsle may have been sexual; Schroder (F 1999) offers a more contextualized reading of the letters.

From Augsburg Mozart and his mother went on to Mannheim, where they remained until the middle of March. Wolfgang became friendly with the Konzertmeister, Christian Cannabich, the Kapellmeister, Ignaz Holzbauer, and the flautist J.B. Wendling; he recommended himself to the elector but with no success. His Mannheim compositions included the keyboard sonatas K309 and 311, the Flute Quartet K285, five accompanied sonatas (K296, K301–3, K305, possibly inspired by the sonatas of Joseph Schuster) and two arias, *Alcandro lo confesso ... Non sò d'onde viene* K294 and *Se al labbro mio non credi ... Il cor dolente* K295; he was also asked by Ferdinand Dejean, an employee of the Dutch East India Company who had worked in eastern Asia for many years as a physician, to compose three flute concertos and two flute quartets, but in the event failed to fulfil the commission and may have written only a single quartet. The aria K294 was composed for Aloysia Lange, the daughter of the Mannheim copyist Fridolin Weber. Mozart, who was in love with Aloysia, put to Leopold the idea of taking her to Italy to become a prima donna, but this proposal infuriated his father, who accused him of dilatoriness, irresponsibility over money and family disloyalty.

In a letter of 11–12 February 1778, Leopold ordered his son to Paris; at this time it was also decided that his mother should continue to accompany him, rather than return to Salzburg, a decision that was to have far-reaching consequences for both father and son. Wolfgang arrived in Paris on 23 March and immediately re-established his acquaintance with Grimm. He composed additional music, mainly choruses (KA1), for a performance of a *Miserere* by Holzbauer and, according to his letters home – which are less than entirely truthful – a sinfonia concertante KAnh.9/297*B*, for flute, oboe, bassoon and horn. Like the *Miserere* choruses, the sinfonia concertante, allegedly suppressed by Joseph Legros, is lost (the convoluted history of this work, and the possibility that part of it survives in KAnh.9/C14.01, is described in Levin, M 1988). A symphony (K297) was performed at the Concert Spirituel on 18 June and repeated several times (as described in his letters, Mozart composed two slow movements, of which the one in 6/8 is probably the original), while a group of ballet pieces, *Les petits riens*, composed for Noverre, was given with Piccinni's opera *Le finte gemelle*.

Mozart was unhappy in Paris: he claimed to have been offered, but to have declined, the post of organist at Versailles, and his letters make it clear that he despised French music and suspected malicious intrigue. He was not paid for a flute and harp concerto (K299) that he had composed in April for the Duc de Guines, and his mother fell ill about mid-June. Although Grimm's doctor was called in to treat her, nothing could be done and she died on 3 July. Mozart wrote to his father to say that she was critically ill, and by the same post to Abbé Bullinger, a close friend in Salzburg, telling him what had happened; Leopold was thus prepared when Bullinger broke the news to him.

These events triggered another round of recriminatory letters: Leopold accused Mozart of indolence, lying and improper attention to his mother; for his part Mozart defended himself as best he could. Although this correspondence is frequently taken to represent the first – and most compelling – evidence of an irreparable fissure in the relationship between Wolfgang and his father, it reflects more on their attempts to come to grips with an overwhelming family tragedy. Leopold's implicit suggestion that Mozart was partly responsible for his mother's death cannot be taken seriously. Stuck in Salzburg, grieving for his wife and worrying about his son, Leopold must have felt himself a helpless bystander; his only recourse was by letter, after the event. Not surprisingly, he sometimes wrote insensitively and hurtfully. His uncompromising devotion to Mozart, however, was never in question. It is significant – given his belief in the fragility of existence (see especially Halliwell, F 1998) – that in his first letter to Wolfgang after learning of Maria Anna's death, he does not lay blame but is concerned chiefly with his son's well-being.

Mozart stayed with Grimm for the remainder of the summer. He had another symphony given at the Concert Spirituel, on 8 September (his claim in a letter of 11 September that it was a new work appears to be untrue), and renewed his acquaintance with J.C. Bach, who had come over from London to hear the Paris singers before composing the opera *Amadis de Gaule*. Mozart also wrote a scena, now lost, for the castrato Tenducci. But his friendship with Grimm, to whom he owed money, deteriorated, and on 31 August Leopold wrote to inform him that, following the death of Adlgasser, a post was open to him in Salzburg, as court organist with accompanying duties rather than as violinist; the archbishop had offered an increase in salary and generous leave. Mozart set out for home on 26 September. Grimm put him on the slow coach through Nancy and Strasbourg to Mannheim, where he heard Benda's melodrama *Medea* and resolved to write one himself (the work, *Semiramis*, if started, was never performed and is now lost; Mozart later wrote a melodrama for the incomplete Singspiel *Zaide*). Leopold, however, was infuriated that Mozart had gone to Mannheim, where, since the removal of Carl Theodor's court to Munich, there were no opportunities for advancement. Mozart reached Munich on 25 December and remained there until 11 January; he was coolly received by Aloysia Weber, now singing in the court opera. Finally, in the third week of January 1779, he arrived back in Salzburg.

Immediately on his return Mozart formally petitioned the archbishop for his new appointment as court organist. His duties included playing in the cathedral, at court and in the chapel, and instructing the choirboys. Reinstated under favourable conditions, he seems at first to have carried out his duties with determination: in 1779–80 he composed the 'Coronation' Mass K317, the *Missa solemnis* K337, the vespers settings K321 and 339 and the *Regina coeli* K276. Nevertheless, Colloredo was not satisfied: in an ambiguously worded document appointing Michael Haydn court and cathedral organist in 1782 he wrote: 'we accordingly appoint [J.M. Haydn] as our court and cathedral organist, in the same fashion as young Mozart was obligated, with the additional stipulation that he show more diligence ... and compose more often for our cathedral and chamber music'. The cause of Colloredo's dissatisfaction may have lain in Mozart's other works of the time: the Concerto for two pianos K365, the Sonata for piano and violin K378, the symphonies K318, 319 and 338, the 'Posthorn' Serenade K320, the Divertimento K334, the Sinfonia concertante for violin and viola K364, incidental music for *Thamos, König in Ägypten* and *Zaide*. Few of these works would have been heard at court, where instrumental music was little favoured, while production of theatrical music was the domain of the civil authorities.

Mozart's contract with Colloredo did not specify his compositional obligations as a composer: it stated only that 'he shall as far as possible serve the court and the church with new compositions made by him'. As Colloredo's criticism makes clear, however, he expected Mozart to take a more active role in the court music. During his final years in Salzburg, then, Mozart reverted to the pattern of 1774–7: his appearances at court as both performer and composer were half-hearted and his music-making was intended instead chiefly for a small circle of friends and the local nobility.

4. The Break with Salzburg and the Early Viennese Years, 1780–83

In the summer of 1780, Mozart received a commission to compose a serious opera for Munich, and the Salzburg cleric Giovanni Battista Varesco was engaged to prepare a libretto based on Danchet's *Idomenée*. The plot concerns King Idomeneus of Crete, who promises Neptune that if spared from a shipwreck he will sacrifice the first person he sees and is met on landing by his son Idamantes. Mozart began to set the text in Salzburg; he already knew several of the singers, from Mannheim, and could draft some of the arias in advance.

Mozart arrived in Munich on 6 November 1780. Both the performing score of the opera (not taken into consideration by the Neue Mozart-Ausgabe; see Münster, J 1982) and Mozart's letters to his father, who was in close touch with Varesco, offer insights into the genesis of the work and its modification during rehearsal. The matters that chiefly occupied Mozart were the need to prune an overlong text, to make the action more natural, and to accommodate the strengths and weaknesses of the singers. Several cuts were made in December, during rehearsals, and Mozart continued to trim the score even after the libretto was sent to the printer at the beginning of January; a second libretto was printed to show the final text (although in the event still more adjustments were made, as the performing score makes clear). Much of the *secco* and accompanied recitative was cut, as well as sections of the ceremonial choral scenes and probably three arias in the last act. In a letter of 15 November to his father, Mozart described his concerns for both dramatic credibility and the singers' capabilities:

> [Raaff] was with me yesterday. I ran through his first aria for him and he was very well pleased with it. Well – the man is old and can no longer show off in such an aria as that in Act 2 – 'Fuor del mar ho un mar nel

> seno'. So, as he has no aria in Act 3 and as his aria in Act 1, owing to the expression of the words, cannot be as cantabile as he would like, he wishes to have a pretty one to sing (instead of the quartet) after his last speech, 'O Creta fortunata! O me felice!' Thus too a useless piece will be got rid of – and Act 3 will be far more effective. In the last scene of Act 2 Idomeneo has an aria or rather a sort of cavatina between the choruses. Here it will be *better* to have a mere recitative, well supported by the instruments. For in this scene which will be the finest in the whole opera ... there will be so much noise and confusion on the stage that an aria at this particular point would cut a poor figure – and moreover there is the thunderstorm, which is not likely to subside during Herr Raaff's aria, is it?

The opera was first given on 29 January 1781, with considerable success. Both Leopold and Nannerl, who had travelled from Salzburg, were in attendance, and the family remained in Munich until mid-March. During this time Mozart composed the recitative and aria *Misera! dove son ... Ah! non son' io che parlo* K369, the Oboe Quartet K370 and possibly three piano sonatas (K330–32, although these may equally date from his first month in Vienna).

On 12 March Mozart was summoned to Vienna, where Archbishop Colloredo and his retinue were temporarily in residence for the celebrations of the accession of Emperor Joseph II; he arrived on 16 March, lodging with the archbishop's entourage. Fresh from his triumphs in Munich, Mozart was offended at being treated like a servant, and the letters that he wrote home over the next three months reflect not only increasing irritation and resentment – on 8 April the archbishop refused to allow him to perform for the emperor at Countess Thun's and thereby earn the equivalent of half his annual Salzburg salary – but also a growing enthusiasm for the possibility of earning his living, at least temporarily, as a freelance in Vienna. Matters came to a head on 9 May: at a stormy interview with Colloredo, Mozart asked for his discharge. At first he was refused, but at a meeting with the chief steward, Count Arco, on 8 June, he was finally and decisively released from Salzburg service, 'with a kick on my arse ... by order of our worthy Prince Archbishop' (letter of 9 June 1781).

About this time Mozart moved to the house of the Webers, his former Mannheim friends, who had relocated to Vienna after Aloysia's marriage to the court actor Joseph Lange, although in order to scotch rumours linking him with the third daughter, Constanze, he relocated again in late August to a room in the Graben. He made a modest living at first, teaching three or four pupils, among them Josepha von Auernhammer (for whom he wrote the Sonata for two pianos K448) and Marie Karoline, Countess Thiennes de Rumbeke, cousin of Count Johann Phillipp von Cobenzl, the court vice-chancellor and chancellor of state (whom Mozart had met in Brussels in autumn 1763). He also participated in, or had works performed at, various concerts: the Tonkünstler-Societät gave one of his symphonies on 3 April (Mozart later applied for membership in the society, which provided pensions and benefits for the widows and orphans of Viennese musicians, but he failed to provide a birth certificate and his application was never approved); and on 23 November he played at a concert sponsored by Johann Michael von Auernhammer. Later Mozart participated in a series of Augarten concerts promoted by Philipp Jakob Martin. At the first of these, on 26 May 1782, he played a two-piano concerto with Josepha von Auernhammer (the programme also included a symphony by him). Mozart's own first public concert took place on 3 March 1782, possibly at the Burgtheater. The programme included the concertos K175 (with the newly composed finale K382) and K415, numbers from *Lucio Silla* and *Idomeneo*, and a free fantasy; on 23 March Mozart wrote to his father that the new concerto finale was 'making ... a furore in Vienna'. During this period he also played regularly at the home of Baron Gottfried van Swieten, where Handel and Bach were staples of the repertory.

By the end of 1781, Mozart had established himself as the finest keyboard player in Vienna – although he was not without competitors, few could match his pianistic feats. The most serious challenge, perhaps, came from Muzio Clementi, with whom Mozart played in an informal contest at Emperor Joseph II's instigation on 24 December. Clearly Mozart was perturbed by the event: although he was judged to have won, and Clementi later spoke generously of his playing, Mozart in his letters repeatedly disparaged the Italian pianist. It is likely that Clementi's skill took Mozart by surprise; the emperor must have been impressed as well, for he continued to speak of the contest for more than a year. That same month saw the appearance of Mozart's first Viennese publication, a set of six keyboard and violin sonatas (K296 and 376–80, of which two, K296 and 378, had been composed earlier). They were well received; a review in C.F. Cramer's *Magazin der Musik* (4 April 1783) described them as 'unique of their kind. Rich in new ideas and traces of their author's great musical genius'.

The most important composition of this period, however, was *Die Entführung aus dem Serail*, the libretto of which was given to Mozart at the end of July 1781. Originally planned for September, the première was postponed until the following summer (Mozart had completed the first act in August 1781). The opera was a great success: Gluck requested an extra performance, Schikaneder's troupe mounted an independent production in September 1784 (although the aria 'Martern aller Arten' was replaced because the orchestra was incapable of performing the obbligato solos), and productions were soon mounted in cities throughout German-speaking Europe. The earliest lengthy obituary of Mozart, in the *Musikalische Korrespondenz der Teutschen Filarmonischen Gesellschaft* of 4 January 1792, described the work as 'the pedestal upon which his fame was erected'.

In his letters to Leopold, Mozart described in detail several of his decisions in composing the opera. He wrote on 26 September 1781:

> in the original libretto Osmin has only [one] short song and nothing else to sing, except in the trio and the finale; so he has been given an aria in Act 1, and he is to have another in Act 2. I have explained to Stephanie the words I require for the aria ['Solche hergelaufne Laffen'] – indeed, I had finished composing most of the music for it before Stephanie knew anything whatever about it. I am enclosing only the beginning and the end, which is bound to have a good effect. Osmin's rage is rendered comical by the use of the Turkish music. In working out the aria I have ... allowed Fischer's beautiful deep notes to glow. The passage 'Drum beim Barte des Propheten' is indeed in the same tempo, but with quick notes; and as Osmin's rage gradually increases, there comes (just when the aria seems to be at an end) the Allegro assai, which is in a totally different metre and in a different key; this is bound to be very effective. For just as a man in such a towering rage oversteps all the bounds of order, moderation and propriety and completely forgets himself, so must the music too forget itself. But since passions, whether violent or not, must never be expressed to the point of exciting disgust, and as music, even in the most terrible situation, must never offend the ear, but must please the listener, or in other words must never cease to be *music*, so I have not chosen a key foreign to F (in which the aria is written) but one related to it – not the nearest, D minor, but the more remote A minor. Let me now turn to Belmonte's aria in A major, 'O wie ängstlich, o wie feurig'. Would you like to know how I have expressed it – and even indicated his throbbing heart? By the two violins playing in octaves. This is the favourite aria of all who have heard it, and it is mine also. I wrote it expressly to suit Adamberger's voice. You see the trembling, the faltering, you see how his throbbing breast begins to swell; this I have expressed by a crescendo. You hear the whispering and the sighing – which I have indicated by the first violins with mutes and a flute playing in unison.

Mozart had already described his concern for naturalness, in both composition and performance, in a letter written in Paris on 12 June 1778:

> Meis[s]ner, as you know, has the bad habit of making his voice tremble at times, turning a note that should be sustained into distinct crotchets, or even quavers – and this I never could endure in him. And really it is a detestable habit and one that is quite contrary to nature. The human voice trembles naturally – but in its own way – and only to such a degree that the effect is beautiful. Such is the nature of the voice; and people imitate it not only on wind instruments, but on string instruments too and even on the keyboard. But the moment the proper limit is overstepped, it is no longer beautiful – because it is contrary to nature.

Shortly after the première of *Die Entführung*, on 16 July, Mozart decided to go forward with his marriage to Constanze Weber (*b* Zell, Wiesental, 5 Jan 1762; *d* Salzburg, 6 Mar 1842), which he had first mooted to his father the previous

December. Events gave him little choice: probably through his future mother-in-law's scheming, he was placed in a position where because of his alleged intimacy with Constanze he was required to agree to marry her or to compensate her. Mozart wrote to his father on 31 July 1782 asking for his approval, on 2 August the couple took communion together, on 3 August the contract was signed, and on 4 August they were married at the Stephansdom. Leopold's grudging consent did not arrive until the next day. The marriage appears to have been a happy one. Although Mozart described Constanze as lacking wit, he also credited her with 'plenty of common sense and the kindest heart in the world', and his letters to her, especially those written when he was on tour in 1789 and when she was taking the cure at Baden in 1791, are full of affection. There is little reason to imagine that she was solely, or even primarily, to blame for their chronic financial troubles, which surfaced only weeks after their marriage; the truth probably lies somewhere nearer Nannerl's statement, in 1792, that Mozart was incapable of managing his own financial affairs and that Constanze was unable to help him. Early 20th-century scholarship severely criticized her as unintelligent, unmusical and even unfaithful, a neglectful and unworthy wife to Mozart. Such assessments (still current) were based on no good evidence, were tainted with anti-feminism and were probably wrong on all counts. But she was an unreliable witness and told many lies about the Requiem, whose completion she had organized: she was of course an interested party. Many of the myths surrounding Mozart's death probably stem from her.

Mozart's departure from Salzburg and his wedding to Constanze set off another acrimonious exchange with Leopold, whose letters from this period are lost (their contents can be inferred from Mozart's). Leopold accused Wolfgang of concealing his affair with Constanze and, worse, of being a dupe, while Wolfgang, for his part, became increasingly anxious to defend his honour against reproaches of improper behaviour and his alleged failure to attend to his religious observations; he chastised his father for withholding consent to his marriage and for his lukewarm reaction to the success of *Die Entführung*. Mozart had reason to be upset: not only had Leopold repeatedly pressed him to return home, but in his dealings with Colloredo Mozart had been told by Count Arco that he could not leave his post without his father's permission. Despite his numerous successes in Vienna, he felt thwarted in his attempt to achieve a well-earned independence.

Presumably in order to heal the rift with his family, Mozart determined to take Constanze to Salzburg to meet his father and sister, although to Leopold's irritation the visit was several times postponed. The success of *Die Entführung* had catapulted Mozart to prominence: the opera was performed at the Burgtheater on 8 October, in the presence of the visiting Russian Grand Duke Paul Petrovich (Mozart directed from the keyboard, as he explained in a letter of 19 October 1782, 'partly to rouse the orchestra, who had gone to sleep a little, partly ... in order to appear before the royal guests as the father of my child'); and between November and March 1783 he played at concerts sponsored by Auernhammer (at the Kärntnertortheater), the Russian Prince Dmitry Golitsïn, Countess Maria Thun, Philipp Jakob Martin (at the casino 'Zur Mehlgrube'), his sister-in-law Aloysia Lange (at the Burgtheater; according to Mozart's letter of 12 March, Gluck, who attended, 'could not praise the symphony and aria too much'), Count Esterházy and the singer Therese Teyber. On 23 March Mozart gave his own academy at the Burgtheater, in the presence of the emperor. The programme may have included the Haffner Symphony K385 (composed in July 1782 to celebrate the ennoblement in Salzburg of Siegmund Haffner) and improvised variations on an aria from Gluck's *La rencontre imprévue*.

Mozart composed several new works for these occasions, including the piano concertos K413–15, later published by Artaria (although Mozart may not have conceived them as a set, the autographs show that some time in the spring of 1783 he thoroughly revised all three together), and three arias, K418–20, intended for a production of Pasquale Anfossi's *Il curioso indiscreto* at the Burgtheater on 30 June 1783. He also began work on the so-called 'Haydn' quartets. The first, K387, was completed in December 1782; the second, K421, was finished in June 1783, while Constanze was giving birth to their first child, Raimund Leopold, born on 17 June.

Mozart and Constanze eventually set out in July (Raimund Leopold, who was left behind, died on 9 August); they remained in Salzburg for about three months. Later correspondence suggests that the visit was not entirely happy – Mozart was anxious about the success of the visit and about his father's reaction to Constanze – but details are lacking. While there, he probably composed his two violin–viola duos for Michael Haydn, who was behindhand with a commission from the archbishop, and parts of the Mass in C minor (K427, never completed) had their first hearing, possibly with Constanze singing, at St Peter's on 26 October. On the return journey to Vienna, Mozart paused at Linz, where he composed a symphony (K425) for a concert; the Piano Sonata K333 may also date from this time.

5. VIENNA, 1784–8

With his return to Vienna in late November 1783, Mozart entered on what were to be the busiest and most successful years of his life. On 22 December he performed a concerto in a concert mounted by the Tonkünstler-Societät, and on 25 January 1784 he conducted a performance of *Die Entführung* for the benefit of Aloysia Lange. He gave three subscription concerts in the private hall of the Trattnerhof in March, and a grand musical academy at the Burgtheater on 1 April; the programme included a 'quite new' symphony, possibly the Linz (K425), a new concerto (K450 or 451), the Quintet for piano and wind (K452) and an improvisation. The 1785 season was similar: there were six subscription concerts at the Mehlgrube beginning on 11 February (including the first performance of the D minor Concerto K466) and another grand academy at the Burgtheater on 10 March. It was chiefly for these concerts that, between February 1784 and December 1786, Mozart composed a dozen piano concertos (from K449 to K503), unquestionably the most important works of their kind. Perhaps in recognition of his risen star, in February 1784 Mozart started keeping a list of his new works, the *Verzeichnüss aller meiner Werke*, recording the incipit and the date of each. The catalogue is a primary source of information concerning Mozart's compositional activities during the 1780s, documenting among other things several lost compositions, including the aria *Ohne Zwang, aus eignem Triebe* K569, the contredanses K565 and an Andante for a violin concerto K470.

In addition to his public performances, Mozart was also in demand for private concerts: in March 1784 alone he played 13 times, mostly at the houses of Count Johann Esterházy and the Russian ambassador, Prince Golitsïn. By the same token, visiting and local virtuosos and concert organizations frequently gave newly commissioned works by him in their programmes: on 23 March the clarinettist Anton Stadler mounted a performance of the Wind Serenade K361, and on 29 April Mozart and the violinist Regina Strinasacchi played the Sonata K454. (Mozart is said to have performed from a blank or fragmentary copy; it is clear from the autograph that the violin part was written first and the piano one added later.) The Tonkünstler-Societät gave the cantata *Davidde penitente* (K469, arranged from the unfinished Mass in C minor K427) in March 1785; Mozart played a concerto for the same group in December. These works and performances brought Mozart considerable acclaim. A review of the December Tonkünstler-Societät concert noted 'the deserved fame of this master, as well known as he is universally valued' (*Wiener Zeitung*, 24 December). Earlier that year Leopold Mozart, who visited Wolfgang in Vienna in February and March 1785, wrote to Nannerl describing a quartet party at Mozart's home at which Haydn told him, 'Before God and as an honest man I tell you that your son is the greatest composer known to me either in person or by name. He has taste and, what is more, the most profound knowledge of composition'.

His publications were numerous. Torricella brought out the three sonatas K333, K284 and K454; in July 1784 Lausch advertised manuscript copies of six piano concertos; and in February 1785 Traeg offered copies of three symphonies. The most significant publications, however, were possibly the three concertos K413–15, published by Artaria in March 1785, and the six quartets dedicated to Haydn, brought out by Artaria in September of that year. The success of these works seems to have brought about a fundamental shift in Mozart's attitude to composition and publishing. After mid-1786, several works were planned primarily with a view to publication rather than public performance; these include the piano quartets K478 and 493, the three piano trios K496, 542 and 548, the C major and G minor string quintets K515 and 516, the Hoffmeister Quartet K499 and the Sonata for piano and violin K526.

Although opera remained central to Mozart's ambitions throughout this period, there was no opportunity to build on the success of *Die Entführung*: by late 1782, Joseph II decided to close down the Nationaltheater (which he had founded in 1776 to promote German-language culture) and to re-establish Italian opera. Mozart was quick to capitalize on the change, although he had little luck in finding a suitable text; on 7 May 1783 he wrote to his father, 'I have looked through at least a hundred librettos and more, but I have scarcely found a single one with which I am satisfied'. He therefore asked Leopold to have Varesco, the Salzburg poet and librettist of *Idomeneo*, provide a text. This was *L'oca del Cairo*, which Mozart received from Salzburg in June 1783. He may have worked on it during his visit to Salzburg, but the project was apparently abandoned by the end of the year, by which time he had sketched out seven pieces, including a large sectional finale. In 1785, or possibly earlier, he began work on *Lo sposo deluso, ossia La rivalità di tre donne per un solo amante*, which he based on the libretto used by Cimarosa for his opera *Le donne rivali* of 1780 (see Zaslaw, in Sadie, B 1996), but this too was left incomplete: of the five surviving numbers – an overture, a quartet, a trio and two arias – only the trio, 'Che accidenti, che tragedia', is completely orchestrated. A one-act comedy, *Der Schauspieldirektor* K486, was given early in 1786 in the Orangerie at Schloss Schönbrunn, together with Salieri's *Prima la musica e poi le parole* (both were commissioned for a visit by the Governor-General of the Austrian Netherlands), and in March a private performance of a revised version of *Idomeneo* was given at Prince Auersperg's; among other changes, Mozart wrote the duet 'Spiegarti non poss'io' (K489) to replace 'S'io non moro a questi accenti' and the scena and rondò 'Non più, tutto ascoltai ... Non temer, amato bene' (K490) to replace the original beginning of Act 2.

The topic of Mozart's first documented collaboration with Lorenzo da Ponte, *Le nozze di Figaro*, was no doubt carefully chosen. Beaumarchais' play, *La folle journée, ou Le mariage de Figaro*, had been printed in German translation in Vienna in 1785, although performances by Schikaneder's theatrical company had been banned; further, it was a sequel to Beaumarchais' *Le barbier de Séville, ou La précaution inutile*, of which Paisiello's operatic version, given at Vienna in May 1784, had been a great success. Work on *Figaro* was started by October or November 1785, and the opera came to the stage of the Burgtheater on 1 May 1786. The initial run was a success: many items were applauded and encored at the first three performances, prompting the emperor to restrict encores at later ones to the arias. Letters from Leopold to Nannerl Mozart make it clear that there was a good deal of intrigue against the work, allegedly by Salieri and Vincenzo Righini, while a pamphlet published in Vienna in 1786 (*Uiber des deutsche Singspiel des Apotheker des Hrn. v. Dittersdorf*; see Eisen, A 1991) similarly claims that '[The foreign partisans] ... have completely lost their wager, for Mozart's *Nozze di Figaro* ... [has] put to shame the ridiculous pride of this fashionable sect'. An equally biting comment appeared in the *Wiener Zeitung* for 11 July: 'Herr Mozart's music was generally admired by connoisseurs already at the first performance, if I except only those whose self-love and conceit will not allow them to find merit in anything not written by themselves'.

The allegedly seditious politics of the opera may be overstated. Da Ponte was careful to remove the more inflammatory elements of Beaumarchais' play, and the characters and events of the opera are well situated within the *commedia dell'arte* tradition. Nevertheless, social tensions remain, as in Figaro's 'Se vuol ballare', the Act 2 finale, and the Count's music early in Act 3. Individual arias also reflect the social standing of the various characters: this may be exemplified by a comparison of Bartolo's blustery, parodistic vengeance aria 'La vendetta' and the Count's 'Vedrò, mentr'io sospiro', with its overtones of power and menace, or between the breadth and smoothness of the Countess's phraseology as opposed to Susanna's. Ultimately, however, *Figaro* may be no more than a comic domestic drama, though not without reflecting contemporary concerns about gender and society (see Hunter, J 1997).

The presumed political implications of Mozart's masonic activities may also be overstated. On 11 December 1784 he had become a freemason at the lodge 'Zur Wohlthätigkeit' ('Beneficence'), which in 1786, at Joseph II's orders, was amalgamated with the lodges 'Zur gekrönten Hoffnung' ('Crowned Hope') and 'Drei Feuern' ('Three Fires') into 'Zur neugrekrönten Hoffnung' ('New Crowned Hope') under the leadership of the well-known scientist Ignaz von Born. The society was essentially one of liberal intellectuals, concerned less with political ideals than with the philosophical ones of the Enlightenment, including nature, reason and the brotherhood of man; the organization was not anti-religious, and membership was compatible with Mozart's faith (Landon, G 1982, suggests that an anonymous oil painting showing a meeting of a Viennese lodge includes, in the lower right corner, a portrait of Mozart). Mozart frequently composed for masonic meetings: the cantata *Die Maurerfreude* K471, for tenor, male chorus and orchestra, was written to honour Born, and various versions of the *Maurerische Trauermusik* K477 were given in 1786 (Autexier, L 1984); several songs and other occasional works, too, were composed for lodge meetings. The masonic style is not restricted to music intended exclusively for lodge performance, but appears elsewhere in Mozart's works, with respect to both general themes, as in *Die Zauberflöte*, and specific musical constructions: Sarastro's aria 'O Isis und Osiris', with its strophic, antiphonal structure, is identical in form with other Viennese masonic songs of the 1780s.

Mozart had first made his way in Vienna by taking pupils, and he continued to do so throughout the mid-1780s: the most important of these was Johann Nepomuk Hummel, who lodged with him between 1786 and 1788. Mozart also taught the English composer Thomas Attwood, whose surviving exercises testify to Mozart's careful, systematic teaching methods, and perhaps carry hints as to how Mozart himself had been taught (see Heartz, H 1973–4). The 'English' connection was already strong at the time of *Figaro*: the first Don Curzio was Michael Kelly (in fact an Irishman), and the first Susanna the soprano Nancy Storace; it is likely that Nancy's brother, Stephen – who later pilfered part of the 'Rondo alla turca' of the Sonata K331 in his opera *The Siege of Belgrade* – also consulted informally with Mozart on matters of composition. (After his return to London, Storace prepared a series of publications which included in 1789 the first edition of the Piano Trio K564, in a text that differs from the first Viennese edition of 1790; he probably received a copy of the work from Mozart himself.)

The impending departure of the English contingent from Vienna, planned for the spring of 1787, led Mozart to consider a journey to London during late 1786, but that idea foundered when Leopold took a strong stand against the proposed trip and refused to look after Mozart's children. (Of Mozart's six children, only two survived to adulthood: Carl Thomas, *b* Vienna, 21 Sept 1784; *d* Milan, 31 Oct 1858, and Franz Xaver, *b* Vienna, 26 July 1791; *d* Carlsbad, 29 July 1844. The others were Raimund Leopold, 1783; Johann Thomas Leopold, 1786; Theresia, 1787–8; Anna Maria, 1789.) Mozart did, however, accept an invitation to Prague, where *Figaro* had been a great success. He spent approximately four weeks there, from 11 January 1787, and clearly relished his popularity in the city. He directed a performance of *Figaro* and gave a concert including a new symphony written for the occasion (the 'Prague', K504 – there is reason to believe that Mozart originally intended to perform the Paris Symphony with a new finale, but, having written it, decided to compose an entirely new symphony altogether; see Tyson, D 1987). And it was at about this time that the Prague impresario Pasquale Bondini commissioned him to write an opera for the following autumn. On his return to Vienna, Mozart asked Da Ponte for another libretto.

The plot of *Don Giovanni*, based like that of *Figaro* on tensions of class and sex, dates back at least to the time of Tirso de Molina (1584–1648), although Da Ponte drew on the most recent stage version, a one-act opera with music by Giuseppe Gazzaniga and a libretto by Giovanni Bertati, given in Venice in February 1787. Mozart left for Prague on 1 October; the première was planned for 14 October 1787, but because of inadequate preparation, *Figaro* was given instead and the new opera was postponed until 29 October, when it earned a warm reception. Mozart directed three or four performances before returning to

Vienna in mid-November. During this time he also visited his friends the Dučeks at their villa outside Prague; he wrote the difficult aria *Bella mia fiamma* K528 for Josefa, an old Salzburg friend. *Don Giovanni* was staged in Vienna in May 1788, with several adaptations: Leporello's escape aria in Act 2 was replaced by a duet with Zerlina; Ottavio's 'Il mio tesoro' in Act 2 was replaced by 'Dalla sua pace' in Act 1, and Elvira was given a magnificent accompanied recitative and aria, 'In quali eccessi ... Mi tradì quell'alma ingrata'.

The two Da Ponte operas, along with the increased success of his publications, initiated a new phase in Mozart's career. Not only did he now give fewer concerts – a grand academy at the Burgtheater on 7 April 1786, less than a month before the première of *Figaro*, was his last in that venue (the programme probably included the C minor Piano Concerto K491) – but other genres came to the fore in his output, including the symphony. The final symphonic triptych, composed between June and August 1788, was apparently intended for a concert series that autumn (Eisen, L 1997); it is striking that Mozart chose these works, rather than concertos, for what may have been his first public concert appearance in two years. Whether these changes were also related to Mozart's appointment the previous December as court *Kammermusicus*, however, is unclear. Apparently he was required to do little more than write dances for court balls; nevertheless, he welcomed the appointment, both for the dependable income it provided and for its advancement of his standing in Viennese musical circles. There is little reason to think that the relatively small salary of 800 gulden (Gluck, the previous incumbent, was paid 2000 gulden) was an insult to Mozart, for the post was superfluous to begin with. Joseph II later remarked that he had created the vacancy solely to keep Mozart in Vienna.

The death of Leopold Mozart in May 1787 may have initiated a fallow period for the composer, albeit at some months' distance. Mozart wrote relatively few works immediately following the Prague première of *Don Giovanni*, among them dances and piano music, songs and arias and at least part of a piano concerto (K537) in addition to the three new items for the Viennese première of his opera. A similar fallow period had followed the death of his mother in Paris in July 1778. Leopold's death also marked the final breakdown of the Salzburg Mozart family. Only Nannerl, who in 1784 had married the magistrate Johann Baptist Franz von Berchtold zu Sonnenburg and moved to St Gilgen, remained, and except for settling their father's estate, Mozart apparently failed to keep in contact with her (his last known letter to her is dated 2 August 1788). Nannerl was hurt by Mozart's lack of attention, so much so that when asked in 1792 to describe his life in Vienna, she pleaded ignorance, despite the fact that she had become personally acquainted with Constanze in 1783 and still had in her possession numerous letters from her father, many of them detailing Mozart's activities at the time.

6. THE FINAL YEARS

Mozart's financial circumstances in Vienna can be measured in part by the locations and sizes of the numerous lodgings he rented there. In January 1784 he moved to the Trattnerhof, and in September of that year to an apartment, now Domgasse 5, in the heart of the town, close to the Stephansdom. By mid-1788, however, he had removed to the distant suburb of Alsergrund, where rents were considerably cheaper. It is from this time that a dismal series of begging letters to his fellow freemason Michael Puchberg survives. One refers to the poor response to his string quintet subscription, another to embarrassing debts to a former landlord, and a third to dealings with a pawnbroker; the letters continued well into 1790.

Mozart's finances during the Vienna years must be counted a mystery. Although he was never forced to do without a maid or other luxuries typical of a person of his standing, his finances were unstable and estimates of his earnings are at best incomplete and unreliable. His main sources of income included profits from his public concerts and payments from private patrons; money earned from teaching; honoraria for publications; and, from 1788, his salary as court *Kammermusicus*. During his early years in Vienna Mozart's performances represented a good source of income. His subscription series of 1784 attracted well over 100 patrons at 6 gulden for three concerts, and, according to Leopold, he took in 559 gulden from his Burgtheater academy on 10 March 1785. He also must have received cash or other rewards from the princes Esterházy and Golitsïn, at whose homes he frequently performed; for his contest with Clementi Joseph II gave him 50 ducats. After 1786, however, this concert-giving income largely disappeared.

Teaching provided less, although Mozart enterprisingly formulated a scheme to ensure some regularity of payment, which he described to his father in a letter of 23 January 1782: 'I no longer charge for 12 lessons, but monthly. I learnt to my cost that my pupils often dropped out for weeks at a time; so now, whether they learn or not, each of them must pay me 6 ducats'. Publications may also have brought in substantial sums, although the payment of 450 gulden that Mozart received from Artaria for the six quartets dedicated to Haydn was exceptional; he received less for the symphonies and the sonatas, quintets and other chamber works printed during the 1780s. On occasion he acted as his own publisher, sometimes with sorry results: a subscription for his string quintets in 1788 apparently failed. In 1791, however, he may have sold copies of *Die Zauberflöte* for 100 gulden each. For the composition of an opera Mozart generally received 450 gulden; payments of this amount are documented for *Die Entführung*, *Figaro* and *La clemenza di Tito* (for *Così fan tutte* see below); his share of the profit from *Die Zauberflöte*, however, is unknown.

Mozart's day-to-day expenses, on the other hand, have been little explored. In addition to rent and food, his income had to cover substantial medical bills (chiefly resulting from Constanze's frequent cures), child-rearing expenses and a costly wardrobe (only one of the prices he paid for maintaining his standing in Viennese society, though gladly it seems). By all accounts he was generous to his friends, sometimes lending them money. Other expenses must be taken into account as well, among them books, music and manuscript paper. Documents show that Mozart was in debt to the publisher Artaria throughout the 1780s, although it is unclear whether this represents monies owed before or after honoraria paid by Artaria for his published works (Ridgewell, G 1999).

The estate documents are difficult to interpret. Mozart was in debt at the time of his death, but not to an excessive degree: the value of his estate, less than 600 gulden, was set against debts of about 900 gulden. However, this does not take into account a judgment of more than 1400 gulden awarded by the courts in November 1791 to Prince Karl Lichnowsky, who had sued Mozart, for unknown reasons (details of the affair and its resolution are known only summarily from an account in the Viennese archives; see Brauneis, G 1991). Nevertheless, Constanze managed not only to pay off Mozart's debts but also to collect the value of the estate. It may be that she was provided for by Mozart's friends and patrons, chief among them van Swieten, or that her finances were secured by the sale of Mozart's music and the income from numerous benefit concerts.

Between 1788 and 1790, van Swieten contributed to Mozart's welfare by having him arrange for private performance several works by Handel, including *Acis and Galatea* (K566, November 1788), *Messiah* (K572, March 1789) and *Alexander's Feast* and the *Ode for St Cecilia's Day* (K591 and 592, both July 1790). But the situation in Vienna at the time was complicated by the Turkish war. One effect of this campaign was a general decline in musical patronage during 1788 and 1789, with fewer concerts than there had been earlier in the 1780s. (The war did provide Mozart with opportunities for composition, however, including the 'Kriegslied' *Ich möchte wohl der Kaiser sein* K539 and the works for mechanical organ, K594, 608 and 616, presumably composed for performance at a mausoleum established in memory of Field Marshal Gideon Laudon, hero of the Siege of Belgrade.)

Perhaps in an effort to alleviate his financial woes, or even to escape what he may have perceived as an oppressive Viennese atmosphere, Mozart undertook a concert tour of Leipzig, Dresden and Berlin in the late spring of 1789. Details of the journey are scarce. At Dresden he played chamber music privately and performed at court, in addition to playing in an informal contest with the organist J.W. Hässler, while at Leipzig he reportedly improvised at the Thomaskirche organ in the presence of the Kantor, J.F. Doles, a former Bach pupil. Mozart may have sold some compositions in Potsdam and Berlin, and he attended a performance of *Die Entführung*. Nevertheless, the journey was not without its rewards. In Leipzig Mozart renewed his acquaintance with Bach's music, obtaining a score of the motet *Singet dem Herrn ein neues Lied!* (BWV225); its impact is evident not only in the chorale of the Armed Men in *Die Zauberflöte* but also, more substantially, in the contrapuntal disposition and character of the finales of his two last string quintets, K593 and 614. And he was probably invited by King Friedrich Wilhelm II, an amateur cellist, to compose quartets and keyboard sonatas. Almost certainly he started work on this commission on the return journey to Vienna: the score of K575 and parts of K589 are written on manuscript paper originating from a mill between Dresden and Prague. When the quartets were finally published by Artaria in 1791, however, they lacked a dedication altogether. Mozart wrote to Puchberg on 12 June 1790, 'I have now been obliged to give away my quartets ... for a pittance, simply in order to have cash in hand'.

His continuing financial problems notwithstanding, Mozart's circumstances were beginning to improve by late 1789. In addition to the first of the 'Prussian' quartets, he wrote two replacement arias for a new production of *Figaro* on 29 August ('Al desio di chi t'adora' K577 and 'Un moto di gioia mi sento' K579, first heard at a Tonkünstler-Societät concert in December), as well as substitute arias for productions of Cimarosa's *I due baroni* (K578), probably for a German-language version of Paisiello's *Il barbiere di Siviglia* (K580), and for Martín y Soler's *Il burbero di buon cuore* (K582 and 583). His work attracted international interest: the poet Friedrich Wilhelm Gotter intended to offer Mozart his opera libretto *Die Geisterinsel* (in the event not set until 1796, by Friedrich Fleischmann), and in April 1791 Mozart was apparently offered a pension by two groups of patrons, one in Amsterdam, the other in Hungary.

His main energies, however, were given to the composition of *Così fan tutte*, his third collaboration with Da Ponte and the only one of the Da Ponte operas for which there is no direct literary source (although, like *Don Giovanni*, it has sources in Tirso de Molina). It may be that the libretto was wholly original to Mozart and the poet, for the subject is sometimes claimed to have been suggested to Mozart and Da Ponte by Joseph II himself, allegedly on the basis of a recent real-life incident. But it is also known that the libretto was initially offered to Salieri, who set some early numbers and then apparently abandoned it (Rice, J 1987). *Così fan tutte* is widely reckoned to be the most carefully and symmetrically constructed of the Da Ponte operas. The three men (the two officers Ferrando and Guglielmo and their friend Don Alfonso) and the three women (the sisters Dorabella and Fiordiligi and their servant Despina) each have an aria in each act; and the ensembles are calculated so that the four principals are kept in their pairs (officers and sisters), and given relatively little personal identity, until well on in Act 2, by which time the sisters are emotionally affected by their disguised lovers. At this point, the pervasive element of parody characteristic of the opera gives way to music more personal in tone, reflecting the characters' differing moral dilemmas.

Little is known of the opera's genesis. It was rehearsed at Mozart's home on 31 December and at the theatre on 21 January 1790 (Puchberg and Haydn probably attended both rehearsals); the première was on 26 January. There were four further performances, then a break because of the death of Joseph II in February, and five more in the summer. Mozart apparently expected to receive 900 gulden for its composition, twice the usual amount, but documents survive only for a payment of 450 gulden (Edge, G 1991). Although the opera was a success – receipts from the court theatre box offices show that it was one of the most heavily attended of the season (Edge, G 1996) – it soon came to be criticized for its apparent moral shortcomings: female fickleness, in particular, was found shocking, and it is made more so by the convention (standing equally in *Figaro* and *Don Giovanni*) that the action should span no more than 24 hours. The opera is susceptible of other interpretations, however. Its appeal to *commedia dell'arte* traditions explains some of the characters and their behaviour, including the use of poison, disguises and elevated rhetoric (Goehring, J 1993), while its balance of sympathy and ridicule presents a commentary on the strength and uncontrollability of amorous feelings and the value of a mature recognition of them.

Joseph II died on 20 February 1790, and with the accession of a new emperor, Leopold II, Mozart hoped for a preferment at court; none was forthcoming. Unlike his predecessor, Leopold (who until his coronation had ruled in Florence as Grand Duke of Tuscany) had musical tastes that were thoroughly Italian. During the two years of his reign he transformed Viennese musical theatre: he planned to replace the old Burgtheater with a magnificent new house, he reintroduced the ballet and revived *opera seria*, and he reformed comic opera. Although these changes were seemingly reactionary, they nevertheless looked to the future: they were responsible at least in part for the composition of *Die Zauberflöte* and *La clemenza di Tito*, both of which were influential in the early 19th century (Rice, J 1995).

In order to take advantage of the coronation festivities, in which he had no official role, Mozart went in September 1790 to Frankfurt, taking his brother-in-law Franz de Paula Hofer and a servant. They arrived on 28 September, and Mozart gave a public concert on 15 October; though musically a success it was poorly attended and financially a failure. On the return journey Mozart gave a concert at Mainz, heard *Figaro* at Mannheim, and played before the King of Naples at Munich. He reached home about 10 November, joining Constanze at their new apartment in central Vienna, to which she had just moved.

A trip to England became a possibility again that autumn. Mozart was tendered an invitation for an opera, but declined (he was also promised an engagement like Haydn's by J.P. Salomon). During the winter months he composed a piano concerto (K595, possibly performed on 9 January 1791 by his pupil Barbara Ployer at a concert held by Prince Adam Auersperg in honour of the visit to Vienna of the King of Naples; see Edge, G 1996) and the last two string quintets (K593 and 614). He played a concerto at a concert organized by the clarinettist Josef Bähr and an aria and a symphony were given at the Tonkünstler-Societät concerts in April. That same month Mozart secured from the city council the reversion to the important and remunerative post of Kapellmeister at the Stephansdom, where the incumbent Leopold Hofmann was aged and ill; he was appointed assistant and deputy, without pay, but in the end Hofmann outlived him.

It was for the festivities at Leopold II's coronation in Prague that Mozart composed *La clemenza di Tito*. Reports published soon after his death suggested that it had been written in only 18 days, some of it in the coach between Vienna and Prague, although it is more likely that it was written over a period of six weeks. The impresario Domenico Guardasoni signed a contract with the Bohemian Estates on 8 July, and his first choice to compose a coronation opera (either on a subject to be suggested by the Grand Burgrave of Bohemia or, if time did not permit, on Metastasio's *La clemenza di Tito*, 1734) was Salieri. But Salieri refused the commission and the work fell to Mozart. Possibly this was in mid-July: the fact that Guardasoni's contract included an 'escape clause', allowing him to engage a different composer, suggests that he may already have expected Salieri to decline and discussed with Mozart the possibility of composing the opera. The text was arranged by Caterino Mazzolà, who cut much of the dialogue and 18 arias while adding four new ones, as well as supplying two duets, three trios and finale ensembles. In his catalogue, Mozart described *Tito* as 'ridotto a vera opera'. The première took place on 6 September.

Mozart's works were widely disseminated in 1791 – Viennese dealers produced nearly a dozen editions of his works in that year alone – and were intended for audiences that ranged far beyond court circles. Among them were the string quintets K593 and 614 (December 1790 and March 1791, respectively), the Concerto K622 for Anton Stadler (for whose basset-clarinet, with its downward extension of a major 3rd, Mozart also probably intended the Quintet K581), the Masonic cantata *Laut verkünde unsre Freude* K623, the aria *Per questa bella mano* K612, the piano variations on *Ein Weib ist das herrlichste Ding* K626, the motet *Ave verum corpus* K618, *Die Zauberflöte* K620 and the Requiem K626. *Die Zauberflöte*, written for Emanuel Schikaneder's suburban Theater auf der Wieden, was well under way by 11 June, as a reference in a letter to Constanze makes clear; possibly it was complete in July except for three vocal items, the overture and the march. The opera has several sources, among them Liebeskind's *Lulu, oder Die Zauberflöte*, published in Wieland's collection of fairy tales, *Dschinnistan* (1786–9); this was a source for other operas given at the Freihaustheater and its rival, the Leopoldstädter-Theater (including Benedikt Schack's *Der Stein der Weisen*, to which Mozart may have contributed several passages in addition to parts of the duet 'Nun, liebes Weibchen, ziehst mit mir' K625; see Buch, K 1997). Many of the ritual elements are derived from Jean Terrasson's novel *Sethos* (1731), which has an ancient Egyptian setting, from contemporary freemasonry and possibly from other theatrical works of the time. The whole belongs firmly in the established traditions of Viennese popular theatre. C.L. Giesecke, a poet, actor and member of the lodge 'Zur neugekrönten Hoffnung', later claimed to be the author of the libretto, but his assertion lacks plausible support. The arguments in favour of Schikaneder's authorship seem incontrovertible.

Although the opera was well received – contemporary opinion on the music was universally favourable – critics found the text unsatisfactory (the *Staats- und gelehrte Zeitung* of Hamburg reported on 14 October that 'the piece would have won universal approval if only the text ... had met minimum expectations'). One hotly disputed point concerns a possible reshaping of the plot while composition was in progress. The opera begins as a traditional tale of a heroic prince (Tamino) rescuing a beautiful princess (Pamina) at the bidding of her mother (the Queen of Night) from her wicked abductor (Sarastro). In the Orator's scene, however, it transpires that the abductor is beneficent and that it is the princess's mother who is wicked. Although it is tempting to think that this shift can only represent a change in plan by Schikaneder and Mozart (traditionally explained as an attempt to avoid duplicating a rival production, Wenzel Müller's *Kaspar der Fagottist, oder Die Zauberzither*), the moral ambiguities that demand explanation if it does not – Sarastro's employment of the evil Monostatos, for example, or the Queen and her Ladies' gifts of the benevolently magical flute and bells to Tamino and Papageno, or Pamina's fear of Sarastro – are not out of line with Viennese popular theatrical traditions, nor with symbolic interpretations of the work. It has also been argued that Tamino's confrontation with the Orator represents a recognition scene, a standard operatic situation also found in *Figaro*, *Don Giovanni* and *Così fan tutte* (Waldoff, J 1994).

Much has been written about freemasonry in the opera. It is unlikely, as has been asserted, that the authors intended the characters to stand for figures involved in the recent history of the movement. They are better understood as generalized and symbolic figures: for instance, Tamino and Pamina are ideal beings seeking self-realization and, especially, ideal union. In this *Die Zauberflöte* may be thought to pursue the theme of self-conscious knowledge predicated in *Così fan tutte*. More broadly, the opera is susceptible to interpretation in the light of the philosophical, cosmological and epistemological background of 18th-century freemasonry as an allegory of 'the quest of the human soul for both inner harmony and enlightenment' (Koenigsberger, J 1975, and Till, J 1992). Such interpretations help to explain how what may superficially seem a mixture of the musically sublime and the textually ridiculous melds into an opera not only theatrically effective but also of a philosophical or religious quality. Goethe tried to write a sequel to it, and Beethoven pointedly quoted from the opera in *Fidelio*.

Probably in mid-July, Mozart was commissioned by Count Walsegg-Stuppach, under conditions of secrecy, to compose a Requiem for his wife, who had died on 14 February 1791; work on this was postponed at least until October 1791, after the completion of *La clemenza di Tito* and *Die Zauberflöte*. It is likely that Mozart was aware of Walsegg's identity: his friend Puchberg lived in Walsegg's Vienna villa, and the inclusion of basset-horns in the score suggests that Mozart could count on the participation of specific players, who would have been booked far in advance for a date and place already known to him. Later sources describe Mozart's feverish work at the Requiem, after his return from Prague, with premonitions of his own death, but these are hard to reconcile with the high spirits of his letters from much of October. Constanze's earliest account, published in Niemetschek's biography of 1798, states that Mozart 'told her of his remarkable request, and at the same time expressed a wish to try his hand at this type of composition, the more so as the higher forms of church music had always appealed to his genius'. There is no hint that the work was a burden to him, as was widely reported in German newspapers from January 1792 onwards.

By the time of Mozart's final illness, he had completed only the 'Requiem aeternam' in its entirety; from the Kyrie to the 'Confutatis', only the vocal parts and basso continuo were fully written out. At the 'Lacrimosa' only the first eight bars are extant for the vocal parts, along with the first two bars for the violins and viola. Sketches for the remaining movements, now mostly lost, probably included vocal parts and basso continuo. Mozart was confined to bed at the end of November; he was attended by the two leading Viennese doctors, Closset and Sallaba, and nursed by Constanze and her youngest sister, Sophie. His condition seemed to improve on 3 December, and the next day his friends Schack, Hofer and the bass F.X. Gerl gathered to sing over with him parts of the unfinished Requiem. He was possibly also visited by Salieri. That evening, however, his condition worsened, and Closset, summoned from the theatre, applied cold compresses; the effect was to send Mozart into shock. He died just before 1 a.m. on 5 December 1791. The cause of his death was registered as 'hitziges Friesel Fieber' (severe miliary fever, where 'miliary' refers to a rash resembling millet-seeds) and later diagnosed as 'rheumatische Entzündungsfieber' (rheumatic inflammatory fever) on evidence from Closset and Sallaba. This seems consistent with the symptoms of Mozart's medical history (Bär, G 1966, 2/1972), more so than various rival diagnoses, such as uraemia (favoured by Greither, G 1970, 3/1977, and Davies, G 1989); there is no credible evidence to support the notion that he was poisoned, by Salieri or anyone else.

Mozart was buried in a common grave, in accordance with contemporary Viennese custom, at the St Marx cemetery outside the city on 7 December. If, as later reports say, no mourners attended, that too is consistent with Viennese burial customs at the time; later Jahn (F 1856) wrote that Salieri, Süssmayr, van Swieten and two other musicians were present. The tale of a storm and snow is false; the day was calm and mild.

7. EARLY WORKS

It is likely that the full extent of Mozart's original output during the 1760s will never be known. Not only were many of his early autographs heavily corrected by his father, but it is clear that some works, such as the pasticcio concertos K37 and 39–41 and to a lesser extent the J.C. Bach arrangements K107, were jointly composed. Other compositions, among them the Sonata for keyboard and violin K8, take over, wholly or in part, movements first written by Leopold. A related problem concerns Leopold's *Verzeichniss* of 1768, which describes '13 symphonies for 2 violins, 2 oboes, 2 horns, viola, and basso, etc.' (Zaslaw, A 1985). Of the early works in the genre attributed to Wolfgang, only eight are demonstrably genuine and known to have been composed by this time, while another four are of uncertain authorship and date. Even if all these symphonies are genuine and early, at least one other is missing. Leopold's list describes additional lost works, including six divertimentos in four parts for various instruments, six trios for two violins and cello, solos for violin and bass, minuets, marches and processionals for trumpets and drums. Also, as with many composers of the time, several works are known only from sources with no direct connection to the composer. Some may be authentic, but in other cases there is insufficient evidence for or against Mozart's authorship (for the symphonies see Eisen, L 1989).

Accounts of Mozart's early stylistic development often fail to take these problems into consideration: demonstrably authentic works are often compared with, and analysed alongside, works only insecurely attributed to Mozart. The inevitable result is a patchwork story of progression and regression. When only the demonstrably authentic works are considered, however, not only does the progression in Mozart's style appear more linear, but individual works, often dismissed as showing no significant evidence of Mozart's development, can be seen to represent new plateaux in his sophistication as a composer. In the case of the symphonies this is especially apparent in the works composed up to about

1771. His earliest works in the genre, composed before 1767, are based on models that he encountered on the 'Grand Tour'. All are in three movements, lacking a minuet and trio, and are scored for two oboes, two horns and strings. The first movements are in expanded binary form, in common time, and have tempo indications of Allegro, Allegro molto or Allegro assai, while the second movements, also in binary form, are in 2/4 time and are marked Andante. The concluding fast movements are generally in rondo form and are marked Allegro assai, Allegro molto or Presto, with 3/8 time signatures. For the most part, these works show a remarkable grasp of the principles of J.C. Bach's symphonic style, including the dramatic contrast of a *forte* motto opening and a *piano* continuation, together with hints of cantabile second subjects. In Vienna in 1768, however, Mozart adopted the common four-movement cycle, as well as local formal preferences: K48, for example, is the first of his symphonies to include a first movement in a fully worked-out sonata form. Still later, in Italy, he reverted to the three-movement pattern with its attendant busy string figuration, lighter textures and less melodic thematic material (but still including full recapitulations, albeit with little or no preceding development). K74, with its linked first two movements, may originally have been intended as an opera overture.

While these symphonies are indebted to models encountered by Mozart during his travels during the 1760s and early 1770s, several depart from local norms in significant ways. The first movement of K16 is an expanded binary form of a type more common among Viennese symphonies; K19 includes a brief diversion based on the dominant minor, a procedure common among Salzburg symphonies of the 1750s; and K22 includes an extended orchestral crescendo and recurrence of tutti primary material at the middle and end of the movement, typical of Mannheim. K112, composed at Milan on 2 November 1771, is unusual for its inclusion of a minuet and trio. This symphony in particular represents a significant advance: it is the first by Mozart to include genuine development, rather than a mere retransition to the recapitulation; it explores a new tonal relationship between minuet and trio (previously always in the subdominant but here in the dominant); and it begins to break down the association, previously strictly upheld, of thematic or motivic material with function. The beginning of the transition, at bar 10, is obscured by a re-use of the symphony's stable opening bar as a jumping-off point for the modulation, an effect heightened by the structure of the opening idea. In earlier symphonies with similarly constructed opening material – an aggressive, *forte* and often unison triadic idea followed by a softer motif characterized by conjunct motion – the first idea is more or less literally repeated; in K112, however, the repetition of the opening is initially lacking and is reserved for the first important cadence, where it serves not only to bring the symmetrical pair of five-bar phrases to a conclusion, but also to represent the first element in a two-bar phrase at the beginning of the transition. This reinterpretation of previously-heard material creates an impression not only of unity, but also of ambiguity, and was to become a standard feature of Mozart's symphonies, and his style in general, during the 1770s and later.

Some departures from local norms may have resulted from Mozart's acquaintance with local Salzburg repertories, which have been underestimated in discussions of his development as a composer of orchestral music. His father was the leading symphonist in the archdiocese, and works by several other composers, including Caspar Christelli, Ferdinand Seidl, Adlgasser and Michael Haydn, were known to Mozart during the 1760s. Many of these include Viennese and Italian features that he encountered at source only later on the 'Grand Tour', as well as novelties of their own. Salzburg also provided Mozart with opportunities for composition: the three serenades K63, 99 and 100 were probably composed there in the summer of 1769. Following local traditions best represented by Leopold Mozart, each has six or more movements plus an associated introductory (and perhaps valedictory) march. More relaxed in style than the symphonies, the serenades show their most refined invention in the slow movements, of which one generally has a concertante part (for violin in K63 and for oboe in K100, which also has concertante parts in a fast movement and the trio of one of its minuets, a pattern that later became standard). The chief influence of Salzburg, however, was on Mozart's church music. The *Missa brevis* K49, although composed in Vienna in 1768, displays all the features of the Salzburg *missa brevis* tradition best represented in the works of Eberlin: in the Kyrie, a slow introduction to the main part of the tutti; solo and tutti writing in the Gloria and Credo, with fugal endings to both; a three-section Sanctus and a solo quartet Benedictus; and a simple, chordal tutti Agnus Dei followed by a lively triple-time 'Dona nobis pacem'. Many other features derive from Italian church music, which was widely disseminated and performed in Salzburg for several decades before the 1760s (Eisen, H 1995). Among these are a preference for da capo arias, which is particularly strong in Mozart's solo church music, including the *Regina coeli* K108, with its large, busy orchestra and soprano solos. The *Litaniae de venerabili altaris sacramento* K125 of 1772 is a more sophisticated and individual work, with strong choral writing, strikingly contrasting arias and an opening Kyrie in an elaborate ritornello structure with three levels – orchestra, chorus and soloists.

In Salzburg, Mozart was also acquainted, both directly and indirectly, with Italian theatrical music even before his numerous tours. Italian operas were often given at court during Schrattenbach's reign, and their style informed the local near-equivalent, the so-called *Finalkomödien*, or school dramas, given annually at the Salzburg Benedictine University to mark the end of the academic year. Mozart composed only one work in this genre, *Apollo et Hyacinthus*, which includes full da capo arias and a striking dialogue for the angry Melia and the innocent Apollo, where changes in texture and key support the sense of drama; it is in many respects a successor to his earlier 'sacred Singspiel' *Die Schuldigkeit des ersten und fürnehmsten Gebots* K35.

La finta semplice, by contrast, gave Mozart his first opportunity to compose *opera buffa*, which required a command of the Italian language, an ability to delineate emotions quickly, a thorough knowledge of a wide range of effective orchestral clichés, and a control of the extended, multi-sectional finales of the Goldoni–Galuppi tradition favoured in Vienna. His next two dramatic works, *Ascanio in Alba* K111 and *Il sogno di Scipione* K126, were of the serenata or *festa teatrale* type. *Ascanio* is a leisurely work, with pastoral choruses and ballets interspersed with the arias, while *Il sogno di Scipione* is less tellingly characterized: the arias are lengthy and contain much bravura writing. The most significant of the early dramatic works, however, is the *opera seria Lucio Silla*, which is less convention-bound and more individual than Mozart's first *opera seria*, *Mitridate, re di Ponto* (modelled in several details of form and treatment on the setting by Quirino Gasparini; see Tagliavini, J 1968). This is particularly true of the role of Junia, whose opening aria alternates an intense Adagio and a fiery Allegro, and whose choral scene at her father's tomb recalls Gluck; the terzetto 'Quell'orgoglioso sdegno', in which the tyrant Sulla expresses his anger, is an early example of simultaneous differentiated characterization. Mozart was clearly pleased with several of the arias, which he had recopied in the later 1770s and early 80s; he may have performed 'Pupille amate' in Vienna as late as Carnival 1786.

8. Works, 1772–81

The pervasive influence of the Italian style lingered on well into the 1770s: it not only informs *La finta giardiniera* and *Il re pastore* but is also found in the church music, including the litanies K195 and 243 (the second of which embraces a variety of styles including simple homophonic choruses as well as dramatic ones, fugues, a plainchant setting and expressive arias with florid embellishment). Several symphonies, among them K181 and 184, are in three movements without a break, on the pattern of the Italian overture, while the A major Symphony K201, composed in April 1774, combines southern grace with an intimate, chamber-music style as well as full-bodied orchestral writing and a Germanic predilection for imitative textures.

No doubt Mozart's interest in counterpoint, as well as a general deepening of his style at this time, was stimulated by his visit in 1773 to Vienna, where he composed six string quartets. (For all its pan-European popularity, the string quartet was little cultivated in Salzburg, where the chief forms of chamber music were the trio for two violins and bass and, during the 1770s, the divertimento for string quartet and two horns. Mozart wrote several such works, including K247 and 287). An altogether more intellectual approach is evident in the quartets: imitative textures are found not only in development sections but in first statements of thematic material as well, while the finales to K168 and K173 are both fugal. Similarly, Mozart's first original keyboard concerto, K175 (possibly intended for organ), exploits counterpoint in ways not previously found in his orchestral music. The finale in particular starts with an imitative gesture that returns in various guises throughout the movement. The Symphony in E♭ K184, its Italianisms notwithstanding, includes a C minor Andante whose main theme is also built on imitation, and the coda to the first movement of the Symphony K201 is a contrapuntal tour de force (the long development section of the finale also includes imitations between basses and first violins). The stormy Viennese style is most apparent in K184 (which was adopted in the 1780s as the overture to T.P. Gebler's *Thamos, König in Ägypten*, for which Mozart also wrote incidental music) and in the G minor Symphony K183. Some of this drama is carried over into the serenades of the mid-1770s, including K185, 203, 204 and 250 (Haffner), which although more relaxed in tone nevertheless frequently touch on a range of affects far beyond those typical of the genre. It was the serenade, in any case, that by 1775 had gained the upper hand in Mozart's orchestral output; there are no Salzburg symphonies – redactions of serenades aside – dating from between 1774 and 1779.

The church music that Mozart composed during this period mostly conforms to Salzburg traditions. The absence of soloists in the Mass K167 recalls Michael Haydn's *Missa S Joannis Nepomuceni* of 1772, while in K275 the distribution of solo and tutti, as well as the contrapuntal endings to the Gloria and Credo, the imitative entries at the beginning of the Sanctus and the solo at the Benedictus are reminiscent of Eberlin. Colloredo's church music reforms, described by Mozart in an oft-cited letter to Padre Martini of 4 September 1776 ('a mass, with the whole Kyrie, the Gloria, the Credo, the epistle sonata, the offertory or motet, the Sanctus and the Agnus, must last no more than three-quarters of an hour'), inform the brevity and style of K192 and 194; both include a minimum of word repetition, simple choral declamation and sparing musical treatment of text meanings, as well as unbroken settings of the Gloria and Credo without extended final fugues. Similar economies are found in K257, 258 and 259. Not all church music composed in Salzburg at this time was subject to Colloredo's reforms, however. A letter written by Leopold Mozart on 1 November 1777 describes a mass by Michael Haydn, the *Missa S Hieronymi*, that lasted an hour and a quarter. And K262 is a long and elaborate work which includes, besides concluding fugues to the Gloria and Credo, contrapuntal writing even at the Kyrie and 'Et incarnatus', and extended orchestral ritornellos.

If the church music mostly fell in step with Salzburg traditions, the symphonies, serenades and concertos of the earlier 1770s differ from other orchestral music composed there not only in their imaginative scoring, formal variety and diverse characters, but also in their susceptibility to critical readings. In the Symphony K133, the opening hammer-strokes do not return at the start of the recapitulation, which begins with the second group, but they appear to be 'realized' in the coda, where the weakly articulated theme first heard in the second bar is repeated with strong, downbeat root motion, reproducing the *forte* dynamic of the hammer-strokes. Not only does this gesture provide stability and closure otherwise lacking in the movement, but there seems little doubt that Mozart considered it quite deliberately. The autograph shows that he originally intended the passage to represent a coda; by cancelling the first ending, however, he integrated it into the movement proper, rather than distancing it from the action. Almost certainly it was works such as this that in Salzburg provoked dissatisfaction with Mozart. For his part, he complained that 'there is no stimulus [there] for my talent. When I play or when any of my compositions is performed, it is just as if the audience were all tables and chairs'.

Shortly before his departure for Paris in autumn 1777 Mozart composed the Piano Concerto K271, which in its scale, mastery of design, virtuosity, elements of surprise (the piano entry in the third bar is unprecedented) and exploitation of the most profound affects, particularly in the recitative sections of the disturbing C minor Andantino, far exceeds his earlier orchestral music. (Some parallels can be found in the violin concertos K216, 218 and 219 of 1775: the first two also have finales in a variety of tempos and metres, while in K219 the soloist is introduced in the first movement by a poetic Adagio episode, and there is a notable 'Turkish' episode in the minuet finale.) In many ways, K271 represents a new, more elaborate style that was to become Mozart's norm in the late 1770s. No doubt personal factors contributed to this development. It is difficult to forgo altogether the notion that the Paris–Mannheim journey of 1777–9, which violently wrenched Mozart from adolescence to manhood, dramatically influenced the style and substance of his music.

Whether as a result of 'foreign' influences or merely a desire to accommodate his works to a specific public, the music that Mozart composed in Mannheim and Paris frequently recalls local styles. Nannerl Mozart remarked of the Piano Sonata K309, written for Christian Cannabich's daughter Rosina, that 'anyone could see it was composed in Mannheim' (letter of 8 December 1777; Leopold, perhaps more astutely, described it on 11 December 1777 as having 'something of the mannered Mannheim style about it, but so little that your own good style is not spoilt'). Nannerl's observation may refer to the sharp dynamic contrasts in the first two movements and the affectation of the Andante; a similar atmosphere is evident in the next sonata, K311. The A minor Sonata K310, on the other hand, follows up the tradition of fiery keyboard writing that Schobert and others had pursued in Paris (although the tripartite Andante cantabile, with its agitated outburst at the centre of the movement, is without expressive precedent). In his six sonatas for keyboard and violin published in Paris (K301–06), Mozart also took over some features of Joseph Schuster's accompanied divertimentos (which he praised in a letter of 6 October 1777 to his father), notably in the structure of the first movement of K303, where the Adagio introduction represents the first subject and recurs at the recapitulation. The sonatas exhibit a wide variety of styles and affects, ranging from the eerie, almost claustrophobic, E minor K304 to the quasi-orchestral K302 (similar variety can be found in the piano sonatas of the mid-1770s, among them the mannered K282 and the orchestral K284). Perhaps the most important orchestral work composed at this time was the Paris Symphony K297. Following Leopold's advice, Mozart carefully tailored the work to local taste, beginning with the obligatory *premier coup d'archet* and continuing with powerful unison and octave passages, brilliant tuttis and exposed passages for the wind. Scored for two flutes, two oboes, two clarinets, two bassoons, two horns, two trumpets, timpani and strings, the symphony consciously exploits the soundscape of the large Paris orchestra.

Formal and textural variety abounds in the works of the mid- to late 1770s. Frequently, as in the Piano Sonata K280, Mozart avoids settling on the dominant (the same process characterizes the Haffner Symphony K385 of 1782), while some works, including the Piano Sonata K311, reverse the order of the material in the recapitulation. Within the recapitulation itself, Mozart finds effective new ways of avoiding a modulation to the dominant, often incorporating further development that relies on earlier transitional material but does not literally duplicate it. A good example is the Paris Symphony, where the introduction of a C♮ in the basses at bar 175 pushes the harmonies to the subdominant side while also, incidentally, serving to disorientate the listener. Because the movement has no internal repeats, the drop to C♮ conjures up memories of the surprising introduction of B♭ at the start of the development, which serves as the jumping-off point for a modulation to the distant key of F major; consequently, on first hearing the recapitulation may seem to represent a 'new' development.

Many of these styles and techniques remained with Mozart after his return to Salzburg in 1779. This is less true of his church music, perhaps, than of his other works, although the Credo of the Coronation Mass K317 has a symphonic thrust lacking in his earlier works and is broken off by an Adagio 'Et incarnatus'; in this respect it shares with Mozart's instrumental compositions of the time a self-conscious exploitation of musical and affective disruption. In the 'Posthorn' Serenade K320, for example, Mozart recalls the striking formal gesture of the Sonata K303, repeating, at the start of the recapitulation, the music of the slow introduction, rewriting it in the prevailing tempo. In the symphonies K318 and 338 Mozart manipulates the recapitulation. K338 repeats only a part of the first theme, reserving the rest for the final cadence, while K318 is altogether novel in its formal outlines, incorporating an Andante after the development and then returning to the second subject before only partly restating the first. Both the Serenade K320 and the magnificent Sinfonia concertante for violin and viola K364 make extensive use of Mannheim-style crescendos. The Andante of K364 in particular represents a peak in Mozart's orchestral style at this time: its rich orchestral textures, with divided violas, verge on the extravagant, while the unwillingness of the soloists to cadence, as they force each other on, often to higher tessituras, gives the movement an almost ecstatic character. (In this regard the Andante is similar in character to the Adagio non troppo of the G minor String Quintet K516, although part of the effect there is harmonic, deriving from the unexpected shifts between minor and major.)

Idomeneo marks the end of this development; it is unquestionably the most complex and opulent work composed by Mozart before his permanent move to Vienna in early 1781. Although nominally an *opera seria*, *Idomeneo* departs substantially from that tradition. With its French source, it is more natural in its expression of emotion and more complex in structure, with a greater emphasis on the participation of the chorus; its scoring, for the virtuoso Mannheim orchestra now at Munich, is exceptionally full and elaborate. The influence of Piccinni's French operas, as well as that of Gluck's reform works, is strong.

A remarkable feature of the opera is its abundance of orchestral recitative, which sharply reflects the sense of the words. It also uses recurrent motifs. Certain phrases recur throughout the opera, referring consistently to individual characters and their predominant emotions, including Ilia's grief, Electra's jealousy and Idamantes' feelings about the sacrifice (Heartz, 'Tonality', J 1974). The key treatment is sometimes unorthodox and invariably expressive, as in Electra's D minor first aria, 'Tutte nel cor vi sento'. Here Mozart reaches a recapitulation in C minor before returning to the home key; he then modulates, without changing speed, into the music of the tempest, also in C minor and making use of a motif similar to that of the aria. The opera's orchestration includes many new and brilliant details, among them the evocative flute, oboe and violin passages in 'Fuor del mar' and the use of sustained wind against inexorable string triplets and muted trumpet fanfares in 'O voto tremendo'. Perhaps the most admired number of the opera is the powerful Act 3 quartet, in which Idamantes resolves to seek death, a tour de force in which intensely chromatic music truthfully embraces four characters' diverse emotions.

9. WORKS, 1781–8

Possibly as a result of the natural development of Mozart's style, or through a wish to accommodate his changed circumstances, the extravagance of Mozart's 'late Salzburg' works gave way, after his permanent move to Vienna, to leaner, more transparent textures and a less ornamental manner. This is true particularly of the six accompanied sonatas published in December 1781 (although only four of them, K376–7, 379 and 380, were composed there; K296 was written at Mannheim, and K378 at Salzburg in 1779 or 1780). At the same time, however, they are broader in conception than the earlier sonatas, with greater forward thrust and, in K380, a deepened sense of rhetorical contrast between full chords and rapid passage-work. Above all, they display a new relationship between the instruments. Although they remain piano sonatas with accompaniment, and contain passages where the violin part could be omitted without damaging the sense of the music, the violin nevertheless increasingly carries essential material, melodic or contrapuntal, and engages in dialogue with the keyboard. The violin part has even greater prominence in K454, composed for Regina Strinasacchi, while in K526, arguably the finest of Mozart's accompanied sonatas, the two instruments are equal in importance. The same trend is evident in the piano trios K496, 502, 542 and 548.

This new equality of partnership is best reflected in the string quartets and quintets of the early to mid-1780s, including the six string quartets dedicated to Haydn, which Mozart described in his dedication of 1 September 1785 as 'the fruits of a long and laborious endeavour', a claim borne out by the relatively large number of quartet fragments from this time as well as by numerous corrections and changes in the autographs (the thorny question of the textual relationship between Mozart's autograph and the first edition, published by Artaria in 1785, is described in Seiffert, N 1997). That Mozart sought to emulate Haydn's quartets op.33, but not to imitate them slavishly, can hardly be doubted: like Haydn's, Mozart's quartets are characterized by textures conceived not merely in four-part harmony, but as four-part discourse, with the actual musical ideas linked to a freshly integrated treatment of the medium. Later critics described them as prime examples, together with those of Haydn and Beethoven, of the 'Classical' quartet, as opposed to the *quatuor concertant* or *quatuor brillant*. According to Koch, they were the finest works of their kind.

Counterpoint in particular takes on a new aspect in the quartets. In the first movements of K421 and 464, each of the principal themes is subjected to imitative treatment; the Andante of K428 follows a similar procedure, supported by increased chromaticism (which is characteristic of the quartet as a whole). The coda of the first movement of the 'Hunt' Quartet K458, like the coda of the earlier A major Symphony K201, draws on the latent imitative potential of the movement's main thematic material, while the famous introduction to the 'Dissonance' Quartet K465 represents an extreme of both free counterpoint and chromaticism. Similar effects can be observed in the C major and G minor quintets of 1787, K515 and 516.

The finale of K387 represents a different use of counterpoint, which is treated not so much as a texture in and of itself, but as a structural topic. Here the main, stable thematic material is represented first and foremost by fugatos, while transitional and cadential material is generally composed in a melody-and-accompaniment *buffo* style. This procedure is reversed in the final movement of the Piano Concerto K459, where fugato represents transition and is explosively elaborated in the double fugue of the central episode. The bifunctionality of Mozart's thematic material in general – as suitable for both counterpoint and melody plus accompaniment – is already adumbrated in the C minor Fugue K426 for two pianos and its later version for strings K546, where the seemingly commonplace Baroque subject erupts at the end of the movement in the previously unimaginable guise of a melody supported by aggressive sawing away in the upper parts. No doubt Mozart had conceived this possibility as early as 1782 while arranging for string quartet several fugues by Bach and Handel: a similar procedure is found at the conclusion of his version of the D♯ minor fugue from Book 2 of Bach's *Das wohltemperirte Clavier*.

The wind music, including the three substantial serenades K361, 375 and 388, shows Mozart's interest in texture in different ways, including the use of novel combinations of instruments (Peter Shaffer, in his play *Amadeus*, puts into Salieri's mouth an evocative description of the opening bars of the Adagio from the Serenade for 13 instruments K361). The C minor Mass K427, meanwhile, includes grave choruses (some in eight parts, as well as the customary four), among which the 'Qui tollis' is built on an ostinato bass of the Baroque descending tetrachord pattern. Several solo items, such as the 'Domine Deus' duet and the 'Quoniam' trio, are almost Handelian in their counterpoint, figuration and bare continuo textures. The Trio for clarinet, viola and piano K498 and the Quintet for piano and wind K452 are both uniquely scored.

Mozart's deliberate attention to even the smallest details of texture, scoring, rhythm and articulation as elements of both affect and style is evident from the numerous erasures, changes and revisions in his autographs. At bar 106 of the first movement of the D minor Piano Concerto K466, for example, he originally wrote the upper string parts as alternating quaver rests and quavers, continuing the pattern of the previous two bars, but he changed these to straight quavers in anticipation of the approaching imperfect cadence. The second movement was initially conceived to begin with the orchestra (as an erased *piano* marking in the first violin part shows) and to include trumpets and drums, and in a possibly related correction, trumpets and drums were omitted from the two final bars of the first movement. In the final movement, at bar 181, Mozart for the first time writes slurs in the accompanying second violin, viola, cello and double bass parts, possibly because their figure here ascends where previously it had descended.

That texture is also a matter of formal significance for Mozart is especially clear in the case of the piano concertos, which have been the subject of continuing debate. H.C. Robbins Landon and others describe the first movements as based on 'pre-classical' ritornello structures inherited from the Baroque and adapted to the newer style, including four tuttis and three solos (Landon, M 1956); current fashion sees them as deriving from the model of the operatic aria (Feldman, M 1996). To some extent, these differing accounts reflect ambiguities in descriptions of the concerto found in 18th-century treatises. Vogler (1779) described the form as identical to the sonata, with the exception that the two parts are not repeated; similar sonata-based derivations are given by Galeazzi (1796) and Kollmann (1799). Scheibe (1745), Quantz (1752), Kirnberger (1771–4) and Türk (1789), however, saw the concerto in terms of ritornello structures. Koch was of two minds: his 1793 *Versuch* identifies a four-tutti form with a third modulating solo, based on the model of C.P.E. Bach; in his 1802 *Lexikon* however, where Mozart is the model, he advocates a three-tutti sonata form model.

The classic formulation is Tovey's, who described Mozart's concertos as a realization of the concerto principle in sonata form (Tovey, H 1936). In this view, the tuttis do not function exclusively, or even primarily, as structural pillars around which concerto movements are built but as contrasting sonorities, reinforcing moments of tonal stability. Through texture and volume, the three main tuttis represent the establishment of the tonic (while at the same time presenting important thematic material and serving as a foil to the later entrance of the piano); the arrival and consolidation of the dominant; and, at the end of the movement, the strongest possible confirmation of the home key and closure. (It is incorrect to describe Mozart's concertos as having a recapitulatory fourth tutti: the material at the return is almost invariably shared between soloist and orchestra; with the possible exception of K467 there are no extended purely orchestral passages comparable in scale to the beginning, medial or final tuttis; and the expressive function of the comingling of tutti and solo functions at the recapitulation is intended to represent rapprochement of the protagonists, not antagonism.) While these principles of construction, in which the character and ordering of the material reflects specific structural concerns, appear to have been worked out first in Mozart's vocal music, the various aria types usually equated with the concerto are traditionally described as 'bithematic ternary' or 'sonata-form' arias; accordingly the two types of works share a basic idea.

Within this larger complex, the first movements of Mozart's concertos follow a pattern consistent in its outlines, and the movements can be divided conveniently into seven large structural units: an opening ritornello including a first theme, extended to a perfect cadence in the tonic, an active half-cadence on the dominant, a more lyrical group (up to *c*1778 this lyrical group tends to appear again in the first solo as the principal theme of the second group while after that date the solo second theme tends to be different) and a concluding group; a first solo that reiterates the first theme followed by an orchestral flourish, confirming the tonality, a modulation to the dominant via new material by the soloist, a flirtation with the stable new key followed by a secondary group, and extension to a perfect cadence in the dominant and a coda; a medial ritornello usually based on one of the forte passages of the opening ritornello; a development-like section representing the first part of the second solo and usually including two parts, the first a drive to a distant key (often the relative minor), the second culminating in a retransition to the tonic; a recapitulation, the second part of the second solo, that largely follows the first solo but omits the modulation; and a concluding ritornello, using material from thde medial ritornello and interrupted by a cadenza. Not only do the specific thematic details and their arrangement vary from work to work, but some of the basic principles of construction are also subject to occasional change: the opening ritornello of K449, for example, is unique among Mozart's concerto first movements for including a modulation (first to the relative minor and then to the dominant where distinctive 'secondary' material is presented), while in K488 the first solo is thematically identical to the opening ritornello and does not include a new theme in the dominant for the soloist (new material is, however, given by the orchestra in the medial ritornello). Second and third movements are less fixed in their structural patterns. *Romance*-type movements, such as that of K466, occur as second movements as do binary types (albeit of considerable complexity), rondos (K491) and variations (K450, 456, 482). Variation movements are also found as finales (K453, 491), as well as sonata forms (K175, original finale, and the violin concerto K207) although the standard pattern is the sonata rondo (notable examples include K271, 456, 459 and 482).

Mozart's contributions to the history of the concerto go beyond formal design. Developments in figuration can already be traced in even the earliest solo concertos (K238 already marks a break from K175 by including a greater variety of left-hand textures), while a noticeable increase in difficulty is apparent in the concertos from 1784 and later (Mozart himself described K450 and 451 as works 'to make the performer sweat' – at least part of the difficulty derives from Mozart's greater simultaneous use of the full range of the keyboard and the ways in which material is split between the hands). Perhaps the most significant development, however, is Mozart's generous accompaniment. The orchestra does not merely accompany *en masse* but also dialogues, sometimes corporately, sometimes individually, both as antagonist and co-protagonist, with the soloist, a trend that is markedly expanded in the concertos from 1784 on; in K482, 488 and 491 the winds achieve parity with the strings as part of the ensemble. This new conception of the relationship between soloist and orchestra is also expressed in Mozart's continuo practice. Unlike in other 18th-century concertos, where the soloist has two functions – as continuo player in the tuttis and soloist elsewhere – Mozart's soloist typically has three: within the large solo sections of the concertos, orchestral outbursts are often accompanied by a soloistic continuo that does not literally duplicate the orchestral basses, projecting a solo personality even in these apparently accompanimental sections. In this respect Mozart's works look forward to concerto styles of the 19th century, where continuo function disappears.

While the model of the early operatic aria is at least partly relevant to Mozart's Viennese concertos, it does not apply to *Die Entführung* or the three Da Ponte operas, *Le nozze di Figaro*, *Don Giovanni* and *Così fan tutte*: by the 1780s Mozart had more or less left earlier aria forms behind (Webster, M 1996). Several different formal types can nevertheless be identified, including binary forms (*Die Entführung*, 'Traurigkeit'), ABA forms (*Don Giovanni*, 'Dalla sua pace', *Così*, 'Un'aura amorosa'), complex two-part forms (*Figaro*, 'Aprite un po' quegl'occhi' and *Don Giovanni*, 'Vedrai, carino'), one-part undivided forms (*Die Entführung*, 'Im Mohrenland'), rondo (in the modern sense; *Così*, 'Donne mie') and rondò (*Figaro*, 'Dove sono'; see Webster, M 1996). In every instance, however, a formal scheme is designed to express the text. The solo arias, rather than representing action, simultaneously portray a variety of complementary or conflicting emotions, one of which usually gains the upper hand. 'Non più andrai' is not so much about Cherubino's implied growth from adolescence to manhood as about Figaro's overwhelming need to gloat; the conflict between achieving peace of mind and inflicting punishment on Belmonte is resolved, in 'O, wie will ich triumphieren', in favour of strangulation; and Don Giovanni's rampant sexual desires overwhelm 'Fin ch'han dal vino', as the final phrase spins, like him, nearly out of control, unable to cadence. Otherwise, the arias often reflect differences in the standing of the various characters – Bartolo's 'La vendetta' is blustery and parodistic, the Count's 'Vedrò, mentre io sospiro', menacing – or express social tension: Figaro's 'Se vuol ballare' is a good example (Allanbrook, J 1982).

The ensembles sometimes carry more complex kinds of expression: the Letter Duet in *Figaro* is a dramatic tour de force, the music representing the dictation of a letter, with phrases realistically repeated and a condensed recapitulation serving for the reading back of the text. But it is the finales in particular that, following *opera buffa* tradition, carry the action forward: changes in tempo, metre, tonality and orchestration resolve existing tensions while creating new ones, always closely allied to the action. Whether they represent meaningful or intentional tonal structures, however, is uncertain. By the same token, the notion that the operas exhibit large-scale tonal planning from start to finish has recently come under attack; many of the key successions cited as evidence of high-level organization are fairly common among Viennese *opere buffe* in general (Platoff, J 1997). In at least parts of some individual operas, however, tonal planning appears to be deliberate. The Act 2 finale to *Don Giovanni*, for example, mirrors almost exactly the tonal action of the opera's overture and *Introduzione*. Both begin in D (minor–major in the overture, major in the finale) and then proceed by way of F (Leporello, Don Giovanni's dance band) to B♭ (Don Giovanni is chased from Anna's bedroom and confronts her father, Elvira confronts Don Giovanni) before returning abruptly to D. The similarity is reinforced by the virtual avoidance of a strong A major in both sections, while the conclusion of the action and the final sextet reverse the minor–major progression of the overture. Strikingly enough, it is the two outer sections of the opera that correspond to the traditional Don Giovanni story; the action 'inside' this frame is the unique contribution of Da Ponte and Mozart.

Shortly after the completion of *Figaro*, and hard on the heels of K503, the last of the concertos composed between 1784 and 1786, came the first of Mozart's 'late' symphonies, the Prague K504. While preserving much of the traditional D major brilliance, this work depends more on the arrangement and development of motifs than on thematic material; its surface is more varied, and more complex, than that of any previous orchestral work by him. The first movement in particular has a structure of great originality. The second-group idea starts as a chromatically inflected variant of the first, with a contrapuntal and sequential continuation, before a distinctive lyrical theme appears, while the development includes contrapuntal workings of various of these motifs and elides with the recapitulation, which fuses the two groups in unexpected ways. The variety of topics and figures alluded to, the integration of learned and *galant* counterpoint, and the rhetorical strategies of the Prague all make it a 'difficult' work, both conceptually and in terms of performance (Sisman, L 1997).

No less difficult are the final three symphonies, K543, 550 and 551, composed in the summer of 1788. K543, like the Prague, includes a long and at times sharply dissonant, tonally wayward introduction, the very sound of which – including clarinets but not oboes – is unprecedented for the time. This was probably the most hastily written of the three: the autograph is among Mozart's most careless, showing numerous mistakes of an elementary sort (instrumental lines are misidentified, necessary clefs and accidentals are omitted, and many parts are written on the wrong staves). More than the G minor or the 'Jupiter', the E♭ major Symphony relies on instrumental doublings, although this, too, contributes to its weighty effect. No less remarkable is the enharmonic writing in the A♭ major Andante con moto, where E♭ is reinterpreted (in bars 92–3) as D♯, leading to an outburst in B minor. Similar enharmonic and chromatic writing is found in the development of the first movement of the G minor

Symphony, which begins with the first-group material in F♯ minor; in the finale, the development begins with a tonally disorientating flourish before embarking on a four-part contrapuntal working-out of the material, ending in the remote key of C♯ minor, where the music pauses before being wrenched back to the tonic for the recapitulation. It is the finale of the 'Jupiter', however, that is best known, although its supposedly 'fugal' writing does not strictly merit that description; rather, it represents an example of *musica combinatoria*, for the various independent motifs heard earlier in the movement are brought together in the coda to create a fugato in five-part invertible counterpoint. In all three of these works, as well as the Prague, the disposition and handling of the orchestra are unique. Building on his experience with concerto and opera, Mozart brought to the symphony orchestra a new understanding of its possibilities both as a corporate body and as a collection of individuals. The textures and gestures range from the most grandiose and 'symphonic' to the most intimate and chamber music-like; the obbligato orchestral ensemble achieves its first perfection in these works.

Mozart's return to the symphony, no doubt related to the increasing prestige of the genre in the mid-1780s, may reflect a fundamental change in his persona as a composer and his ideas of self-presentation. The final triptych forms a natural conclusion, both stylistically and biographically, to this period. But it is also fair to identify a similarly fundamental change in the works composed from 1784 onwards: beginning with the Concerto K450, Mozart's music is significantly more complex, more expansive, larger in scale and more difficult than previously (that Mozart himself may have been in some way aware of this is documented perhaps by the thematic catalogue of his works that he began at this time). This change is apparent from a comparison between the earlier three of the six quartets dedicated to Haydn, written in 1782–3, and the later three, written in 1784–5. Similarly, the Concerto K449, completed in February 1784 but, as the autograph shows, probably begun over a year earlier, is stylistically more akin to the less ambitious early Viennese concertos (K413–15) than to its successors.

During the 19th century, this division of Mozart's works into two stylistic phases, the first up to the end of 1783, the second from 1784 onwards (a division tacitly recognized by theorists, who predominantly cite the later works), fused with then current biographical views of the composer as a divinely inspired genius – by implication a paragon of balance, regularity, symmetry and logic – to endorse a view of the 'Classical style', and Mozart's relationship to it, that has persisted in writings on the composer until the end of the 20th century. As a result, several anomalous works, chief among them the final three symphonies and the C minor Concerto K491, are sometimes seen as representing a social rebellion, a 'critical world view', or Mozart's disillusionment with the Viennese musical public (see McClary, M 1986, Kerman, M 1991, and Subotnik,

J 1991–2, but in light of Powers, H 1995). It is just as valid, however, to see these works as assertions of self-awareness. Mozart's plays of wit and his elaborate musical sophistication are not restricted to a handful of works: the abrupt shift from B♭ major to B minor in the central episode of the finale of the Concerto K456 or the precipitous modulation from B♭ to F♯ minor in the first movement of the Trio K563, the introduction of new themes in the development sections of the quartets K458 and 464, the three simultaneous dances in the Act 1 finale of *Don Giovanni* and the over-elaborate, almost decadent, ornamentation in the slow movement of the Concerto K450 all testify to a style that in general is concerned less with thematic unity and regularity than with disjunction and surprise. The final apotheosis of the 'Jupiter' does not represent a revelation of the symphony's teleological goal, nor is it a comment on the social 'norms' implied by that formulation. Rather, it signifies a self-realization of 'the intellectual force that activates the structure of the work ... that side-steps the coherence of form' (Chua, L 1999). In this respect, it is not wayward, but typical of Mozart's music of the mid-1780s.

10. WORKS, 1789–91

The Clarinet Quintet K581 of September 1789 is a late manifestation of the 'Classical' style of the mid-1780s, and in particular of Mozart's ability to create and weld together a diversity of gestures over the course of entire paragraphs and entire movements. This is most notably the case at the arrival on the dominant in the first movement: a rest in all the parts – more a signal to stop the action after a tutti arrival than an indication of any particular length of silence – is followed by a pizzicato cello line outlining the tonic and fifth of the harmony, long held notes in the second violin and viola that seem almost to emerge from the preceding silence and a new lyrical melody in the first violin. The re-entry of the clarinet with the same melody signals further changes: a shift to the minor mode, lower dynamics and syncopations in the strings. All of these lead to a confrontation between the clarinet and the rest of the ensemble, an outbreak of semiquavers and a conclusive trill, on three instruments, resulting in the firmest cadence in the movement to that point. The effect is to drag the listener along on a wave of increasingly agitated activity; in this respect it resembles the successively more elaborate waves of pianistic activity that animate the first solo of the Concerto K467.

Yet the Clarinet Quintet is not generally representative of Mozart's prevailing style at the time, which is often characterized as ironic, restrained or serenely detached. Some commentators date the origin of this style to the time of the last three symphonies, others to that of *Don Giovanni* or even the two string quintets of 1787. No doubt there are similar elements in other works of the period 1784–8: the Concerto K503 is sometimes described as neutral or cold. But on the whole the late works can be characterized as noticeably more austere and refined than the earlier works, more motivic and contrapuntal, more economical in the use of material and texturally less rich. There are fewer new themes in development sections or in exposition codas, and second-group material is frequently derived from primary ones by some form of extension or contrapuntal treatment.

This is particularly true of the late quintets K593 and 614. K593 has a first movement in a style more spare in texture than that of the preceding quintets but polyphonically richer, most obviously in the recapitulation, where the exposition material is extended and elaborated. The same can be said of K614, the minuet of which is canonic; more impressive still is the finale, the development section of which includes a double fugue. At the same time, both quintets self-consciously exploit similar topics – each first-movement Allegro begins with a passage imitating horns, while that of K614 retains something of a wind serenade atmosphere – while making use of textures in novel ways. The Adagio of K593, not unlike the slow movement of the G minor Quintet K516, is a study in sonorities: each of its five large paragraphs is similarly structured around a recurring pattern, beginning with the full ensemble, reducing to three parts (the violins and viola alternating with the violas and cello) and then returning to five. K614 is novel in a different way. Here the first movement can be seen as a contest between the first violin and the rest of the ensemble, achieving rapprochement only in the final bars. (A similar principle is in evidence in the Piano Trio K502, where the exposition, development and recapitulation each represent an increasingly complex dialogue between piano and violin, with the cello fully participatory only after the second theme.) The textures of the late quartets, however, seem tame by comparison. Mozart must have realized that the new, elaborately wrought four-part quartet style he had previously cultivated would not serve for the concertante quartets popular in Berlin, and for the last two movements of K589 and the last three of K590, presumably conceived in the first instance for the cello-playing King of Prussia, the idea of the cello's prominence seems virtually to have been abandoned. It may also be that hopes of a preferment there – or of successfully completing the commission – had faded.

The notion of a contest in the first movement of K614 suggests that play on genre, consisting in this case of tension between the brilliant and 'Classical' styles identified by early writers on string chamber music, is also self-consciously present in Mozart's works of the late 1780s (it had been there earlier, as well, in the Piano and Wind Quintet K452, a concerto in all but name, and in the final movement of the Piano Sonata K333, which includes a concerto cadenza). But there is a twist: in some instances Mozart manipulates not merely markers of genre, but markers of form and procedure as well. The slow movement of the E♭ Quintet K614, ostensibly a theme and variations (and among the most popular of Mozart's late variation sets, as several contemporaneous arrangements for keyboard show), takes over characteristic gestures of the rondo (including tonic restatements of the main theme) and, more importantly, the sonata. The passages linking the variations are typical sonata transitions, while the climax of the movement, which includes some of the sharpest dissonances in all of Mozart, corresponds to the increase in harmonic tension characteristic of a sonata development. A clear return to both tonic and main theme characterizes the final variation, which is followed by a sonata-like coda, drawing together the main procedural gestures of the movement.

Mozart's interest in Baroque counterpoint, so evident in the late quintets, may have been rekindled by his Handel arrangements for van Swieten and his trip in 1789 to Leipzig, where he renewed his acquaintance with Bach's works. Although the influence of Bach had been strong during the early 1780s, when Mozart also transcribed several preludes and fugues for van Swieten, a truly Classical, integrated counterpoint of a Bachian sort appears to have become a regular feature of his music only in the late 1780s. Sometimes the counterpoint is explicit, as in the central fugato of the overture to *Die Zauberflöte* or in the chorale of the Men in Armour in that opera; for the most part, however, it is subsumed within larger forms and textures. In the Variations K613 the introduction and the theme, the song *Ein Weib ist das herrlichste Ding*, are combined contrapuntally in the coda, while in the Piano Sonata K576 the main secondary material of both outer movements is contrapuntally derived from the primary material (the first movement also includes significant contrapuntal working in the development and recapitulation).

Chance dictated that Mozart, in his last months, should compose works in three genres with which he had been little occupied for almost a decade: the Singspiel *Die Zauberflöte*, the Requiem and the *opera seria La clemenza di Tito*. Until the 1960s Mozart scholars were inclined to dismiss *Tito* as an opera written hastily and with distaste. Yet there is no reason to imagine that Mozart had reservations about composing it; serious opera had always attracted him, and many composers were setting Metastasio's classical librettos modified to meet contemporary taste through the addition of ensembles and choruses. Certainly the opera is written in a style more austere than that of the Da Ponte operas, but it is appropriate to the topic. It is clear that the aria lengths were carefully planned. In Act 2, both the prima donna (Vitellia) and the primo uomo (Sextus) have full-length rondò arias; Sextus's arias involve progressive increases of tempo, no doubt intended to represent the screwing up of his courage. The arias for the other characters, including Titus, are much shorter, while the trios embody some degree of simultaneous representation of different emotions, as in the *opere buffe*.

The Act 1 finale, however, moves in a sense opposite from that of the traditional, accelerating *opera buffa* ensemble of confusion. It starts Allegro and ends Andante, with the principals on stage bewailing the betrayal of Titus while the groans of the populace are heard in the distance.

Die Zauberflöte and the Requiem appear on first hearing to be dramatically different in conception – no work by Mozart is more heterogeneous or displays as broad a range of stylistic references as the opera, while the Requiem seems to refer uniquely to its own rarefied spiritual domain – yet both exploit contrast to an extreme. The opera's fugal overture, with its key of E♭ and three introductory chords, is symbolically masonic; other ritual music, including Sarastro's songs, the choruses and some of the ensembles, also derives from freemasonry. Papageno's strophic comic songs, on the other hand, are in the cheerful manner of other contemporary Singspiele. The songs for the serious characters, while rarely using the extended forms of Italian opera, are more italianate; among these are Tamino's lyrical Portrait Aria and the Queen of Night's two bravura arias. Pamina's lament, 'Ach, ich fühl's', falls in between. Its simple, intimate manner reflects her more universal, idealized character. The remarkable Orator's Scene in the Act 1 finale, however, is sui generis (while at the same time recalling Mozart's interest in declaimed musical settings, first evident in the late 1770s).

The Requiem, by contrast, hides its diversity. Nevertheless the three prevailing textures – homophonic or chordal as in the 'Dies irae' and 'Rex tremendae', contrapuntal as in the 'Requiem aeternam', the Kyrie fugue and the 'Recordare', and cantabile as in the 'Te decet hymnus' and 'Tuba mirum' – are juxtaposed almost kaleidoscopically, often succeeding each other in response to single phrases of the text. At times, the enharmonic and chromatic modulations are extreme, notably in the 'Confutatis' (from bar 25), where the successive lines of text are given in A minor, A♭ minor, G minor and then, via F♯ major, F major (Wolff, I 1991). The make-up of the ensemble, including basset-horns, bassoons, trumpets, timpani and strings (with obbligato trombone in the 'Tuba mirum'), but no flutes, oboes or horns, lends itself to an extraordinarily beautiful, dark-hued sound. In the 'Rex tremendae' and in particular the 'Confutatis', the orchestra represents a character in its own right.

11. AFTERMATH: RECEPTION AND SCHOLARSHIP

To judge by the more than normally laudatory tone of the obituaries and other tributes, Mozart's reputation stood high at the time of his death; although his music was frequently criticized as too audacious and complex, it was understood that he was an artist far out of the ordinary. In 1795, the

Teutschlands Annalen des Jahres 1794 reported that 'In this year ... nothing can or may be sung or played, and nothing heard with approbation, but that it bears on its brow the all-powerful and magic name of Mozart', and by the end of the century his music held centre stage across Europe. Many of the mature works were already well known during the 1780s: the six quartets dedicated to Haydn, published by Artaria in September 1785, were available in Paris as early as December of that year, and some piano concertos were performed regularly in London from January 1786 onwards. It was *Die Entführung*, however, that first established Mozart's fame and influence throughout German-speaking Europe. The opera had been given in more than 20 cities by 1786, and Goethe, in his *Italienische Reise* of 1787, wrote that 'All our endeavours ... to confine ourselves to what is simple and limited were lost when Mozart appeared. *Die Entführung aus dem Serail* conquered all.' Most of the other mature operas were similarly well received. Both *Figaro* and *Don Giovanni* were widely performed, especially in German, while *Così* had received numerous performances by 1793; *Die Zauberflöte* was universally popular. *La clemenza di Tito*, on the other hand, was slower to gain public acceptance (except in England, where it remained the favoured Mozart opera until the second decade of the 19th century).

No doubt interest in Mozart's music was fuelled by his premature death and by stories concerning the Requiem that began circulating shortly afterwards. The earliest known account, published in the *Salzburger Intelligenzblatt* for 7 January 1792, already adumbrated what is by now a familiar tale:

> Some months before his death he received an unsigned letter, asking him to write a requiem and to ask for it what he wanted. Because this work did not at all appeal to him, he thought, I will ask for so much that the patron will certainly leave me alone. A servant came the next day for his answer. Mozart informed the unknown patron that he could not write it for less than 60 ducats and then not before two or three months. The servant returned immediately with 30 ducats and said that he would ask again in three months and that if the mass were ready, he would immediately hand over the other half of the money. So Mozart had to write it, which he did, often with tears in his eyes, constantly saying: I fear that I am writing a requiem for myself.

This anecdote neatly summarizes the Romantic image of Mozart that was prevalent throughout the 19th century and much of the 20th, although numerous documented facts and other evidence contradict it. Mozart may have fallen ill as early as his visit to Prague in September 1791, but there is no sign of any protracted bad health that could have given rise to increasingly dark thoughts about his mortality and the work he was engaged on. Nor did the Requiem exclusively occupy his time: both the Clarinet Concerto K622 and the masonic cantata *Laut verkünde unsre Freude* K623 were completed in the autumn. Circumstantial evidence suggests that Mozart probably knew more about the commission than has generally been supposed.

In view of the specific details of the anecdote, which are of a sort unlikely to have been known to the general public so soon after Mozart's death, it may have originated with Mozart's inner circle: from the beginning, apparently, someone was determined to cast Mozart's life in a particular, and not entirely truthful, light (although see Clarke, I 1996). It was only a small step from this first fabrication to a web of stories intended to promote various myths about the composer: that he was an 'eternal child', a social rebel, a libertine, a misunderstood genius, a helpless victim of professional conspiracies, or even an idiot savant who cared for nothing but his music (for a good summary, see Stafford, G 1991). Much of the Mozart myth, including his alleged poverty and neglect in Vienna, as well as the jealousy of rival composers, was in place by 1800, when Thomas Busby wrote in the *Monthly Magazine* (London, December 1798):

> Had not the almost uniform practice of courts long explained to mankind the principle on which they act, how difficult would it be to conceive, that that of Vienna could so little appreciate the merit of this extraordinary man, who looked to it for an asylum, and passed in its vicinity the last ten years of his life! the dispensers of royal favours, whose ears imbibe with such avidity the flattery that meanness offers, can neglect that genius which nobly refuses the tale of adulation; can stifle it with poverty, and even follow it with persecution.

Contradictory as the numerous biographical tropes surrounding the composer's life may at first seem, they nevertheless add up to a remarkably consistent picture of Mozart as an artist and personality distinctly outside the 'norm'. And it was this notion of Mozart's lack of connection to the real world that set a course for Mozart scholarship – whether biographical, analytical or editorial – up to the end of the 20th century.

Even the earliest biographies took sides in the struggle to present an 'authentic' version of Mozart's life: Nannerl's account, dealing mostly with the Salzburg years, is included in the obituary of Friedrich Schlichtegroll (F 1793), while Constanze's position was first put forward by Niemetschek (F 1798); it is worth noting that Constanze bought up and destroyed the entire edition of the publication containing Schlichtegroll's obituary, apparently disliking its portrayal of her. A more substantial presentation of this side of the story is the biography by Georg Nikolaus Nissen, Constanze's second husband (F 1828), which served as the main source for many later accounts, including those of Oulibicheff (Ulïbïshev) (F 1843) and Holmes (F 1845) (the year after the publication of Nissen's biography Vincent and Mary Novello met Constanze and Nannerl, both of whom talked about Mozart; see Medici and Hughes, G 1955). The first important scholarly biography, embodying fresh research, appeared in the centenary year, 1856 – Otto Jahn's *W.A. Mozart* (F 1856). Ludwig von Köchel's chronological thematic catalogue of Mozart's works, ahead of its time in scholarly method, appeared six years later.

In the early decades of the 20th century, Mozart scholarship was dominated by Wyzewa and Saint-Foix's highly schematic analytical and stylistic study of the works (F 1912–46); Alfred Einstein, in particular, took over many of their conclusions in his edition, the third, of the Köchel catalogue (1937). Similarly important are Dent's pioneering study of the operas (J 1913), Schiedermair's presentation of the letters (A 1914) and Hermann Abert's revision of Jahn (F 1919–21). Emily Anderson's edition of the letters, with revised editions appearing in 1966 and 1985, was published in 1938 (Anderson, A 1938); although it remains the fullest English translation available, it has been superseded by the complete German edition of W.A. Bauer, O.E. Deutsch and J.H. Eibl (A 1962–75). The sixth edition of the Köchel catalogue, published in 1964, included substantial new information but by the late 1990s was badly out of date; a more reliable guide to the authenticity, chronology, history and sources for Mozart's works is found in the prefaces and critical reports to the Neue Mozart-Ausgabe (1955–91). The known documents relating to Mozart's life and works are collected in Deutsch's *Mozart: die Dokumente seines Lebens* (A 1961, with supplements in 1978, 1991 and 1997).

Despite the dramatic increase in Mozart research in the late 20th century, and the renewed availability of numerous sources since the recovery in Poland of autographs lost during World War II, modern scholarship continues to rely on a limited range of material. This is especially evident in editions of Mozart's works, which are based almost exclusively on the autographs, for the most part ignoring, or at least undervaluing, contemporaneous manuscript copies and printed editions. This editorial stance has as much to do with past perceptions of Mozart as with modern notions of textual scholarship: the idea that his works were in some way 'perfect', and that transmission inevitably involves corruption, resulted in a misunderstanding of the essential nature of autographs as representing performance and in the dismissal of some sources that were considered less important, including even Mozart's own performing copies. By the same token, the study of the autographs themselves was for many years limited by a Mozart-centred outlook. Between 1800, when the Offenbach publisher J.A. André purchased the bulk of Mozart's estate from Constanze, and the 1960s and 70s, when Wolfgang Plath published his important articles on *Schriftchronologie*, interest in these documents centred chiefly on the identification and chronological development of Mozart's handwriting. It was only in the 1970s that the watermarks began to be taken into account, in Alan Tyson's systematic and pioneering study, which gave rise to substantial revisions in the dating of Mozart's works. Since then, source studies have broadened in scope to include not only contemporaneous copies, but also Mozart's sketches (Konrad, E 1992) and first editions of his works (Haberkamp, A 1986). Nevertheless, much remains to be done.

Analytical studies in the 1980s and 90s also departed from traditional formal and Schenkerian models (although these have remained vital). Contextual, topical, rhetorical and genre- and gender-based studies have become prominent, not only in the operas but also in Mozart's instrumental music, chiefly the symphonies and concertos. These two orchestral genres in particular lie at the heart of performing practice studies, an important element of Mozart scholarship from the 1970s onwards. Biography, finally, has continued to command attention, displaying a wide range of concerns from the psychological (Hildesheimer, F 1977, and Solomon, F 1995, but see also Head, F 1999) to the increasingly important contextual (Braunbehrens, F 1986, Halliwell, F 1998).

Works

Editions: *Wolfgang Amadeus Mozart's Werke*, ed. J. Brahms and others (Leipzig, 1877–1905/*R*) [MW]
Wolfgang Amadeus Mozart: Neue Ausgabe sämtlicher Werke, ed. Internationale Stiftung Mozarteum Salzburg (Kassel, 1955–91) [NMA; nos. shown, e.g. Series (IV): Werkgruppe (3)/Abteilung (2)/Band (i), page (273) – IV:3/2/i, 273; Abteilung and Band nos. not always applicable]

Thematic catalogue: L. von Köchel: *Chronologisch-thematisches Verzeichnis sämtlicher Tonwerke Wolfgang Amade Mozarts* (Leipzig, 1862; rev. 2/1905 by P. Graf von Waldersee; rev. 3/1937 by A. Einstein, repr. with suppl. 1947; rev. 6/1964/*R* by F. Giegling, A. Weinmann and G. Sievers)

K – *no. in Köchel, 1862; for items not in 1862 edn, no. from 2/1905 or 3/1937 given*
K⁶ – *no. in Köchel, 6/1964; nos. preceded by A, B or C in appendices*
KMS – *nos. in Konrad, E1992*
A – *Anhang* [appx]*: applicable only to edns of Köchel before 6/1964*
BH – *no. in Breitkopf edn*
LC – *Leopold Mozart's catalogue, 17??; see Zaslaw, A1985*
(D) – *date from MS of work (not always clear)*
(L) – *date from Mozart's letters*
(V) – *date from Mozart: Verzeichnüss aller meiner Werke* (1784–91), in *GB-Lbl*

Editions published in Mozart's lifetime are noted in the Remarks column, excluding arrangements and, generally, pf reductions; references to movements are shown in small roman, e.g. K320/iii.

Items are arranged in each category by order of K⁶ numbers

MASSES, MASS MOVEMENTS, REQUIEM

K	K[6]	*Title*	*Key*	*Scoring*	*Composition*	*MW*	*NMA*	*Remarks*	
33	33	Kyrie	F	SATB, str, bc	Paris, 12 June 1766 (D)	III/i, 2	I:1/1/vi, 3		
139	47*a*	Missa solemnis	c	S, A, T, B, SATB, 2 ob, 2 tpt, timp, str, bc	? Vienna, aut. 1768	I/i, 117	I:1/1/i, 37	'Waisenhausmesse'; perf. orphanage in Rennweg, Vienna, 7 Dec 1768	8
49	47*d*	Missa brevis	G	S, A, T, B, SATB, str, bc	Vienna, Oct–Nov 1768 (D)	I/i, 1	I:1/1/i, 3	sketch, KMS 1768[a] KA20*a*/636*b*, 25	38
65	61*a*	Missa brevis	d	S, A, T, B, SATB, str, bc	Salzburg, 14 Jan 1769 (D)	I/i, 33	I:1/1/i, 159	perf. Salzburg, collegiate church, 5 Feb 1769; KMS 1769[a]	
66	66	Missa	C	S, A, T, B, SATB, 2 ob, [2 hn,] 4 tpt, timp, str	Salzburg, Oct 1769 (D)	I/i, 49	I:1/1/i, 185	'Dominicus' Mass; perf. Salzburg, St Peter, 15 Oct 1769, for Cajetan Hagenauer; hn parts *c*1775–6; KMS 1769[a]	8
89	73*k*	Kyrie	G	SSSSS (? or soloistic)	Salzburg, 1772	III/ii	I:1/1/vi, 6	KMS 1772[a]	
90	90	Kyrie	d	SATB, bc	Salzburg, 1772	—	I:1/1/vi, 13		
167	167	Missa	C	SATB, 2 ob, 4 tpt, 2 vn, b, bc	Salzburg, June 1773 (D)	I/i, 179	I:1/1/ii, 3	'In honorem Ssmae Trinitatis'	13, 40
192	186*f*	Missa brevis	F	S, A, T, B, SATB, [2 tpt,] 2 vn, b, bc	Salzburg, 24 June 1774 (D)	I/i, 239	I:1/1/ii, 75	tpt parts added later	13, 40
194	186*h*	Missa brevis	D	S, A, T, B, SATB, 2 vn, b, bc	Salzburg, 8 Aug 1774 (D)	I/i, 265	I:1/1/ii, 121		13, 40
220	196*b*	Missa brevis	C	S, A, T, B, SATB, 2 tpt, timp, 2 vn, b, bc	? Salzburg, 1775–7	I/i, 291	I:1/1/ii, 163	'Spatzenmesse'	14
262	246*a*	Missa [longa]	C	S, A, T, B, SATB, 2 ob, 2 hn, 2 tpt, 2 vn, b, bc	Salzburg, 1775	I/ii, 119	I:1/1/ii, 197	2 tpt added *c*1777	14, 40
257	257	Missa	C	S, A, T, B, SATB, 2 ob, 2 tpt, timp, 2 vn, b, bc	Salzburg, Nov 1776 (D)	I/ii, 1	I:1/1/iii, 3	'Credo'; KMS 1776[a]	14, 40
258	258	Missa brevis	C	S, A, T, B, SATB, 2 ob, 2 tpt, timp, 2 vn, b, bc	Salzburg, Dec ?1775 [?1776] (D)	I/ii, 55	I:1/1/iii, 115	'Spaur', but possibly not mass composed for consecration of Count Friedrich Franz Joseph von Spaur, Feb 1777	14, 40
259	259	Missa brevis	C	S, A, T, B, SATB, 2 ob, 2 tpt, timp, 2 vn, b, bc	Salzburg, Dec 1776 (D)	I/ii, 89	I:1/1/iii, 195	'Organ solo'; 2 ob added ? 1776–81	14, 40
275	272*b*	Missa brevis	B♭	S, A, T, B, SATB, 2 vn, b, bc	Salzburg, by 1780	I/ii, 183	I:1/1/iv, 3	perf. Salzburg, St Peter, 21 Dec 1777	14, 40
317	317	Missa	C	S, A, T, B, SATB, 2 ob, 2 hn, 2 tpt, timp, 2 vn, b, org	Salzburg, 23 March 1779 (D)	I/ii, 207	I:1/1/iv, 57	'Coronation'	17, 42

K	K6	Title	Key	Scoring	Composition	MW	NMA	Remarks	
337	337	Missa	C	S, A, T, B, SATB, 2 ob, 2 bn, 2 tpt, timp, 2 vn, b, bc	Salzburg, March 1780 (D)	I/ii, 255	I:1/1/iv, 193	autograph incl. rejected 136-bar frag. Cr	17
341	368*a*	Kyrie	d	SATB, 2 fl, 2 ob, 2 cl, 2 bn, 4 hn, 2 tpt, timp, str, bc	? Munich, 1780–81, or ? Vienna, late 1780s	III/i, 31	I:1/1/vi, 84	lacks authentic sources	
427	417*a*	Missa	c	2 S, SATB, 2 basset-hn, 2 bn, 2 tpt, timp, str, bc	Vienna, *c*July 1782, ? Salzburg, Oct 1783	XXIV, no.29	I:1/1/v	Cr inc., Ag not composed; Ky, Gl, San perf. Salzburg, St Peter, 26 Oct 1783; see Davidde penitente K469; KMS 1782[b, c, d1–5], 1783	23, 45
626	626	Requiem	d	S, A, T, B, SATB, 2 basset-hn, 2 bn, 2 tpt, timp, str, bc	Vienna, late 1791	XXIV, no.1	I:1/2/i–ii	inc.; completed by F.X. Süssmayr and others	33, 35, 53 54, 55

Frags: K223/166*e*, Osanna, C, 21 bars, 1772, NMA, I:1/1/vi, 15; KA18/166*f*, Ky, C, 49 bars, 1772, NMA, I:1/1/vi, 17; KA19/166*g*, Ky, D, 12 bars, 1772, NMA, I:1/1/vi, 29; KA16/196*a*, Ky, G, 34 bars, ?1787–90, NMA, I:1/1/vi, 46; KA13/258*a*, Ky, C, 9 bars, ?1790–91, NMA, I:1/1/vi, 82; KA322/296*a* = KA12/296*b*, Ky, E♭, 34 bars, ? early 1778, MW, III/1, 11, NMA, I:1/1/vi, 31; KA12/296*c*, San, C, 21 bars, ?1779–80, NMA, I:1/1/vi, 80; KA15/323, Ky, C, 37 bars, ?1787–90, MW, III/i, 22, NMA, I:1/1/vi, 50; KA20/323*a*, Gl, C, 26 bars, ?1787–90, NMA, I:1/1/vi, 76; KA14/422*a*, Ky, D, 11 bars, ?1787–90, NMA, I:1/1/vi, 80

Doubtful (selective list): K—/KA[3a]235*f*/C1.02, Mass, E♭, by B. Schack, in Periodical Collection of Sacred Music no.4 (London, 1831), 'additions by Mozart'; KA234/C1.08, Mass, C, numerous sources attrib. W.A. Mozart, pubd as Duae missae, no.1 (Munich), and as Novello no.8; K—/C1.18, 'Missa solemnis pastorita', G (Munich, 1946); K—/C1.20, Missa solemnis, C, also attrib. F. Brixi, Bs = K92/K[3]92/C3.01; K140/K[3]235*d*/C1.12, Missa brevis, G, unattrib. parts *D-Abk* with autograph corrections, NMA, I:1/1/i, 285; K340/K[3]186*f*/C3.06, Ky, C, lost, MS copy once owned by J.A. André

Spurious (selective list): K115/166*d*, Missa solemnis, ed. R. Kubik (Neuhausen-Stuttgart, 1981), vs ed. D. Townsend (New York, 1963), frag. draft *D-OF*, ed. in MW, XXIV, no.28 and by W. Schulze (Stuttgart, 1983), by L. Mozart (S4.2), see Pfannhauser (D 1971–2); K91/186*i*, Ky, by G. Reutter (ii); K221/A1, Ky, MW, XXIV, no.34, by J.E. Eberlin; K116+—/90*a*+417*B*+A18–19, Missa brevis, MW, XXIV, no.33, by L. Mozart, see Plath and others, D 1971–2; KA233/C1.06, Novello no.7, by F.X. Süssmayr according to C. Mozart, attrib. Pichler at *A-GÖ*

LITANIES, VESPERS, VESPER PSALMS

K	K6	Title	Key	Scoring	Composition	MW	NMA	
109	74*e*	Litaniae lauretanae BVM	B♭	S, A, T, B, SATB, 2 vn, b, bc	Salzburg, May 1771 (D)	II, 1	I:2/i, 13	11
125	125	Litaniae de venerabili altaris sacramento	B♭	S, A, T, B, SATB, 2 ob/fl, 2 hn, 2 tpt, str, bc	Salzburg, March 1772 (D)	II, 13	I:2/i, 23	13, 38
195	186*d*	Litaniae lauretanae BVM	D	S, A, T, B, SATB, 2 ob, 2 hn, str, bc	Salzburg, May 1774 (D)	II, 63	I:2/i, 135	13, 39
193	186*g*	Dixit Dominus, Magnificat	C	S, T, SATB, 2 tpt, timp, 2 vn, b, bc	Salzburg, July 1774 (D)	II, 169	I:2/ii, 1	
243	243	Litaniae de venerabili altaris sacramento	E♭	S, A, T, B, SATB, 2 ob/fl, 2 bn, 2 hn, str, bc	Salzburg, March 1776 (D)	II, 109	I:2/i, 251	14, 39
321	321	Vesperae de Dominica	C	S, A, T, B, SATB, 2 tpt, timp, 2 vn, b, bc	Salzburg, 1779 (D)	II, 193	I:2/ii, 33	17
339	339	Vesperae solennes de confessore	C	S, A, T, B, SATB, 2 tpt, timp, 2 vn, b, bc	Salzburg, 1780 (D)	II, 237	I:2/ii, 101	17

Frag.: K—/321*a*, Mag, C, 7 bars, NMA, I:2/ii, 18

SHORT SACRED WORKS

K	K⁶	Title	Key	Scoring	Composition	MW	NMA	Remarks	
20	20	God is our Refuge	g	SATB	London, July 1765 (D)	III/i, 47	III:9, 2	motet; autograph (partly L. Mozart) given to *GB-Lbl*, July 1765, see King, *Festschrift Albi Rosenthal*, ed. R. Elvers (Tutzing, 1984), 157–80	
—	33*c*	Stabat mater		SATB	? by 1768	—	—	lost; in LC	
—	41*f*	[Fugue à 4 voci]		?SATB	? by 1768	—	—	lost; in LC	
47	47	Veni Sancte Spiritus	C	S, A, T, B, SATB, 2 ob, 2 hn, 2 tpt, timp, bc	? before 1770	III/i, 48	I:3, 12	traditionally considered identical to Veni in LC, Vienna, 1768	
—	47*b*				Vienna, late 1768	—	—	lost; 'grand offertory' perf. Vienna, Waisenhauskirche, 7 Dec 1768, ? = K34	8
117	66*a*	Benedictus sit Deus	C	S, SATB, 2 fl, 2 hn, 2 tpt, timp, str, bc	Salzburg, ?1769	III/ii, 21	I:3, 25		8
141	66*b*	Te Deum	C	SATB, 4 tpt, timp, 2 vn, b, bc	?Salzburg	II/i, 133	I:3, 43	earliest source 1770s; orig. timp part lost	8
85	73*s*	Miserere	a	ATB, b	Bologna, July–Aug 1770 (D)	III/i, 58	I:3, 69	last 3 verses ?incorrectly attrib. J. André in one MS	
86	73*v*	Quaerite primum	d	SATB	Bologna, 9/10 Oct 1770	III/i, 62	I:3, 73	ant; exercise for Accademia Filarmonica, Bologna; copies in *A-Sm*, *I-Baf* transmit version by G.B. Martini, *I-Bc*	9
108	74*d*	Regina coeli	C	S, SATB, 2 ob/fl, 2 hn, 2 tpt, timp, str, bc	Salzburg, May 1771 (D)	III/i, 63	I:3, 74		10, 38
72	74*f*	Inter natos mulierum	G	SATB, 2 vn, b, bc	?Salzburg	III/ii, 9	I:3, 9	off, for feast of St John the Baptist, 24 June; traditionally dated 1771 but earliest source late 1770s	
127	127	Regina coeli	B♭	S, SATB, 2 ob/fl, 2 hn, str, bc	Salzburg, May 1772 (D)	III/i, 87	I:3, 120		13
143	73*a*	Ergo interest	G	S, str, bc	Milan or Salzburg, 1772–3	III/ii, 37	I:3, 62	motet	
165	158*a*	Exsultate, jubilate	F	S, 2 ob, 2 hn, str, bc	Milan, Jan 1773 (D)	III/ii, 43	I:3, 157	motet, for V. Rauzzini; perf. Milan, 17 Jan 1773; rev. version with 2 fl in place of 2 ob, text changes, Salzburg, about 1780: see R. Münster, *Mozart-Studien*, ii (1993), 119–33	11
222	205*a*	Misericordias Domini	d	SATB, 2 vn, [va,] b, bc	Munich, early 1775	III/ii, 77	I:3, 182	off	
260	248*a*	Venite populi	D	SSAATTBB, 2 vn ad lib, b, bc	Salzburg, 1776 (D)	III/ii, 91	I:3, 199	off	
277	277	Alma Dei creatoris	F	S, A, T, B, SATB, 2 vn, b, bc	Salzburg, by 1781	III/ii, 111	I:3, 223	off	14

K	K^6	Title	Key	Scoring	Composition	MW	NMA	Remarks	
273	273	Sancta Maria, mater Dei	F	SATB, str, bc	Salzburg, 9 Sept 1777 (D)	III/ii, 103	I:3, 234	grad, for feast of BVM, 12 Sept; ? for St Peter, Salzburg	
A1	297*a*	Miserere (8 movts)		SATB, orch	Paris, March–April 1778	—	—	for work by I. Holzbauer; lost; see letter, 5 April 1778	
146	317*b*	Kommet her, ihr frechen Sünder	B♭	S, str, bc	Salzburg, 1770s	VI/i, 81	I:4/iv, 33	aria, usually said to date from 1779	
276	321*b*	Regina coeli	C	S, A, T, B, SATB, 2 ob, 2 tpt, timp, 2 vn, b, bc	Salzburg, late 1770s	III/i, 118	I:3, 243	lacks authentic sources	17
343	336*c*	O Gottes Lamm; Als aus Aegypten	F; C	S, bc	? Vienna or Prague, ? 1787–8	III/i, 154	III:8, 30	Ger. sacred songs (see letter, 29 May 1787 and first edition)	
618	618	Ave verum corpus	D	SATB, str, bc	Baden, 17 June 1791 (D)	III/ii, 123	I:3, 261	motet	33

Frag.: KA23/166*b*, In te Domine speravi, C, 34 bars, 1774

Doubtful: K34, Scande coeli limina, off, C, ? Kloster Seeon, Bavaria, early 1767, MW, III/ii, 1, NMA, I:3m 3, ? = K—/47*b*; K142/C3.04, Tantum ergo, B♭, ? by J. Zach; K—, Amen, MW, III/i, 144, NMA, I:3, 270, see Eisen, D 1991, 271–2; K197/C3.05, Tantum ergo, D, MW, III/i, 149, NMA, I:3, 276, transmitted with version, attrib. Mozart, of K142/C3.04

Spurious: K44/73*u*, Musica super cantum gregorianum, by J. Stadlmayr, see 'Arrangements'; KA21/$K^3$93*c*/A2, Lacrimosa, K326/$K^3$93*c*/A4, hymn, both by Eberlin; KA238/A17, Stabat mater, by P.E.F. Ligniville; K177/C3.09, off, by L. Mozart

CHURCH SONATAS

K	K^6	Key	Scoring	Composition	MW	NMA
67	41*b*	E♭	2 vn, b, org	Salzburg, late 1771–1772	XXIII, 1	VI: 16, 2
68	41*i*	B♭	2 vn, b, org	Salzburg, late 1771–1772	XXIII, 3	VI: 16, 4
69	41*k*	D	2 vn, b, org	Salzburg, late 1771–1772	XXIII, 5	VI: 16, 6
144	124*a*	D	2 vn, b, org	Salzburg, early 1774	XXIII, 7	VI: 16, 8
145	124*b*	F	2 vn, b, org	Salzburg, early 1774	XXIII, 9	VI: 16, 11
212	212	B♭	2 vn, b, org	Salzburg, July 1775 (D)	XXIII, 11	VI: 16, 13
241	241	G	2 vn, b, org	Salzburg, Jan 1776 (D)	—	VI: 16, 16
224	241*a*	F	2 vn, b, org	Salzburg, 1779–80	XXIII, 14	VI: 16, 18
225	241*b*	A	2 vn, b, org	Salzburg, 1779–80	XXIII, 18	VI: 16, 22
244	244	F	2 vn, b, org [solo]	Salzburg, April 1776 (D)	XXIII, 21	VI: 16, 25
245	245	D	2 vn, b, org [solo]	Salzburg, April 1776 (D)	XXIII, 24	VI: 16, 28
263	263	C	2 tpt, 2 vn, b, org [solo]	Salzburg, late 1776	—	VI: 16, 32
274	271*d*	G	2 vn, b, org	Salzburg, 1777 (D)	XXIII, 27	VI: 16, 36
278	271*e*	C	2 ob, 2 tpt, timp, 2 vn, b, org	Salzburg, March–April 1777	XXIII, 30	VI: 16, 39

K	K[6]	Key	Scoring	Composition	MW	NMA
329	317*a*	C	2 ob, 2 hn, 2 tpt, timp, 2 vn, b, org [solo]	Salzburg, early 1779	XXIII, 41	VI: 16, 49
328	317*c*	C	2 vn, b, org [solo]	Salzburg, early 1779	XXIII, 36	VI: 16, 60
336	336*d*	C	2 vn, b, org [solo]	Salzburg, March 1780 (D)	XXIII, 51	VI: 16, 65

Spurious: KA65*a*/124*A*, by L. Mozart

ORATORIOS, SACRED DRAMAS, CANTATAS

K	K[6]	*Title (description, libretto)*	*Scoring*	*Composition*	*MW*	*NMA*	*Remarks*	
35	35	Die Schuldigkeit des ersten und fürnehmsten Gebots (pt 1 of orat, 3, I.A. Weiser)	3 S, 2 T, 2 ob/fl, 2 bn, 2 hn, trbn, str	Salzburg, early 1767	V/i	I:4/i	perf. Salzburg, 12 March 1767; pt 2 by J.M. Haydn, pt 3 by A.C. Adlgasser	7, 39
42	35*a*	Grabmusik (cant.)	S, B, SATB, [2 ob,] 2 hn, str	Salzburg, 1767	IV/1, 1	I:4/iv, 1	? perf. Salzburg Cathedral, 7 April 1767; final recit and chorus added *c*1773	7
118	74*c*	La Betulia liberata (orat, 2, P. Metastasio)	4 S, T, B, SATB, 2 ob/fl, 2 bn, 4 hn, 2 tpt, str	Italy and Salzburg, March–July 1771	IV/2, 1	I:4/ii	commissioned in Padua, apparently unperf.	10
469	469	Davidde penitente (orat, 2, ? L. Da Ponte)	2 S, T, SATB, 2 fl, 2 ob, 2 cl, 2 bn, 2 hn, 3 trbn, timp, str	Vienna, March 1785	IV/2, 1	I:4/iii	music from Mass K427/417*a* except for 2 arias, 6 and 11 March 1785 (V); perf. Vienna, Burg, 13 March	24
471	471	Die Maurerfreude (cant., F. Petran)	T, TTB, 2 ob, cl, 2 hn, str	Vienna, 20 April 1785 (V)	IV/1, 24	I:4/iv, 35	perf. Vienna, lodge 'Zur gekrönten Hoffnung', 24 April 1785 (Vienna, 1785)	26
619	619	Die ihr des unermesslichen Weltalls Schöpfer ehrt (cant., F.H. Ziegenhagen)	S, pf	Vienna, July 1791 (V)	VII/1, 82	I:4/iv, 59	sketch in autograph	
623	623	Laut verkünde unsre Freude (cant., E. Schikaneder)	2 T, B, fl, 2 ob, 2 hn, str	Vienna, 15 Nov 1791 (V)	IV/1, 40	I:4/iv, 65	perf. Vienna, lodge 'Zur neugekrönten Hoffnung', 17 Nov 1791	33, 56

Frag.: K429/468*a*, Dir, Seele des Weltalls (cant., L.L. Haschka), T, TTB, fl, 2 ob, cl, 2 hn, bn, str, Vienna, 1785–6, MW, XXIV, no.36a–b, NMA, I:4/iv, 96, partly completed by M. Stadler

Spurious: K623/623*a*, Lasst uns mit geschlungnen Händen, S, ?, appended to 1st edn of K623 (Vienna, 1792)

OPERAS, MUSICAL PLAYS, DRAMATIC CANTATAS

K	K⁶	Title (description, libretto)	Scoring	First performed	MW	NMA	Remarks	
38	38	Apollo et Hyacinthus (Lat. int, 3, R. Widl)	2 S, 2 A, T, B, 2 ob, 2 hn, str	Salzburg, Benedictine University, 13 May 1767	V/ii	II:5/i	perf. with Widl's Lat. play, *Clementia Croesi*	7
51	46*a*	La finta semplice (ob, 3, C. Goldoni, rev. M. Coltellini)	3 S, 2 T, 2 B, 2 fl/eng hn, 2 ob, 2 bn, 2 hn, str	Salzburg, Archbishop's Palace, on or about 1 May 1769	V/iv	II:5/ii	composed Vienna, mid-1768	8, 39
50	46*b*	Bastien und Bastienne (Spl, 1, F.W. Weiskern, J. Müller and J.A. Schachtner, after M.-J.-B. Favart and H. de Guerville: *Les amours de Bastien et Bastienne*)	S, T, B, 2 ob/fl, 2 hn, str	Vienna, F.A. Mesmer's house, ?Sept–Oct 1768	V/iii	II:5/iii	see Tyler J 1990	8
87	74*a*	Mitridate, re di Ponto (dramma per musica, 3, V.A. Cigna-Santi after G. Parini's It. trans. of J. Racine: *Mithridate*)	4 S, A, 2 T, 2 fl, 2 ob, 2 bn, 4 hn, str	Milan, Regio Ducal, 26 Dec 1770	V/v	II:5/iv	aria 'Vado incontro al fato estremo' (Act 3 scene iii) by Q. Gasparini (see Peiretti, J 1996); KMS 1770$^{\alpha-\varepsilon}$	9, 39
111	111	Ascanio in Alba (festa teatrale, 2, G. Parini)	4 S, T, SATB, 2 fl, 2 cb/eng hn/serpentini, 2 bn, 2 hn, 2 tpt/hn, timp, str	Milan, Regio Ducal, 17 Oct 1771	V/vi	II:5/v	for wedding of Archduke Ferdinand of Austria and Maria Beatrice Ricciarda of Modena, with ballet KA207/C27.06	11, 39
126	126	Il sogno di Scipione (azione teatrale, 1, Metastasio)	2 S, 3 T, 2 fl, 2 ob, 2 bn, 2 hn, 2 tpt, timp, str	? Salzburg, Archbishop's Palace, May 1772	V/vii	II:5/vi	composed ?April–Aug 1771, ? given as serenata at enthronement of Count H. Colloredo as Prince-Archbishop of Salzburg	14, 39
135	135	Lucio Silla (dramma per musica, 3, G. De Gamerra)	4 S, 2 T, SATB, 2 ob/fl, 2 bn, 2 hn, 2 tpt, timp, str	Milan, Regio Ducal, 26 Dec 1772	V/viii	II:5/vii		10, 11, 20, 39
196	196	La finta giardiniera (ob, 3, ? G. Petrosellini)	4 S, 2 T, B, 2 fl, 2 ob, 2 bn, 2 hn, 2 tpt/hn, timp, str	Munich, Salvator, 13 Jan 1775	V/ix	II:5/viii	perf. as Spl, Die verstellte Gärtnerin, Augsburg, 1 May 1780; autograph Act 1 lost	13, 39
208	208	Il re pastore (serenata, 2, Metastasio)	3 S, 2 T, 2 fl, 2 ob/eng hn, 2 bn, 2 hn, 2 tpt/hn, str	Salzburg, Archbishop's Palace, 23 April 1775	V/x	II:5/ix		13, 39
A11	315*e*	Semiramis (duodrama, O. von Gemmingen)	2 hn, 2 tpt/hn, str	Mannheim, Nov 1778 (L)	—	—	lost, ? never begun	
345	336*a*	Thamos, König in Ägypten (play with music, 5, T.P. Gebler)	B, SATB, 2 fl, 2 ob, 2 bn, 2 hn, 2 tpt, 3 trbn, timp, str	Salzburg, ?1773 and ?1776–9	V/xii	II:6/i	? 2 choruses composed Vienna, 1773; final version ?1776–9	12, 17, 40
344	336*b*	Zaide (Das Serail) (Spl, 2, Schachtner, after F.J. Sebastiani: *Das Serail*)	S, 2/3 T, 2 B, 2 fl, 2 ob, 2 bn, 2 hn, 2 tpt, timp, str	Frankfurt, 27 Jan 1866	V/xi	II:5/x	Composed Salzburg, 1780, inc.; lacks ov. and final chorus; KMS 1779[a]	17
366	366	Idomeneo, re di Creta (dramma per musica, 3, G.B. Varesco, after A. Danchet: *Idomenée*)	3 S, 3 T, B, SATB, 2 fl, 2 ob, 2 cl, 2 bn, 2 hn, 2 tpt, timp, str	(i) Munich, Residenz, 29 Jan 1781	V/xiii	II:5/xi	with ballet, K367	18, 19, 20, 25, 43
				(ii) Vienna, Palais Auersperg, 13 March 1786			perf. with K489, 490, both composed by 10 March 1786 (V)	

K	K[6]	Title (description, libretto)	Scoring	First performed	MW	NMA	Remarks	
384	384	Die Entführung aus dem Serail (Spl, 3, C.F. Bretzner: *Belmont und Constanze*, rev. G. Stephanie the younger)	2 S, 2 T, B, SATB, 2 fl/pic, 2 ob, 2 cl/basset-hn, 2 bn, 2 hn, 2 tpt, timp, str	Vienna, Burg, 16 July 1782	V/xv	II:5/xii	vs (Mainz, 1785–6) KMS 1781[a]	20, 22, 29, 48, 55
422	422	L'oca del Cairo (ob, 2, Varesco)	3 S, 2 T, 2 B, [chorus,] 2 ob, 2 bn, 2 hn, str	unperf.	XXIV, no.37	II:5/xiii	composed Salzburg and Vienna, late 1783, inc.; 1 trio completed, 6 nos. sketched	25
430	424*a*	Lo sposo deluso (ob, 2, after *Le donne rivali*; attrib. ? Petrosellini)	2 S, 2 T, 2 fl, 2 ob, 2 bn, 2 hn, 2 tpt, timp, str	unperf.	XXIV, no.38	II:5/xiv	begun ?1785; only ov., trio and qt completed; KMS 1783[5a, 6, γ]; A. Campana, *MJb 1988–9*, 573–88	25
486	486	Der Schauspieldirektor (Spl, 1, Stephanie the younger)	2 S, T, B, 2 fl, 2 ob, 2 cl, 2 bn, 2 hn, 2 tpt, timp, str	Schloss Schönbrunn, Orangery, 7 Feb 1786	V/xvi	II:5/xv	completed Vienna, 3 Feb 1786 (V), perf. with A. Salieri's Prima la musica; KMS 1785[a], 1786[γ/1–2]	25
492	492	Le nozze di Figaro (ob, 4, Da Ponte, after P.-A. Beaumarchais: *La folle journée, ou Le mariage de Figaro*)	5 S, 1/2 T, 3/4 B, SATB, 2 fl, 2 ob, 2 cl, 2 bn, 2 hn, 2 tpt, timp, str	(i) Vienna, Burg, 1 May 1786	V/xvii	II:5/xvi	completed Vienna, 29 April 1786 (V); vs (Bonn, 1795); numerous sketches	25, 26, 27, 29, 48, 49, 55
				(ii) Vienna, Burg, 29 Aug 1789			with arias K577, 579	
527	527	Il dissoluto punito, ossia Il Don Giovanni (ob, 2, Da Ponte)	3 S, T, 4 B, SATB, 2 fl, 2 ob, 2 cl, 2 bn, 2 hn, 2 tpt, 3 trbn, timp, mand, str	(i) Prague, National, 29 Oct 1787	V/xviii	II:5/xvii (concert version of ov., IV: 11/x, 23)	Prague, 28 Oct 1787 (V); vs (Mainz, 1791) and (Vienna, 1790–91); KMS, 1787[b]	27, 28, 48, 49, 51, 55
				(ii) Vienna, Burg, 7 May 1788			perf. with addns K540*a*, *b*, *c*	
588	588	Così fan tutte, ossia La scuola degli amanti (ob, 2, Da Ponte)	3 S, T, 2 B, SATB, 2 fl, 2 ob, 2 cl, 2 bn, 2 hn, 2 tpt, timp, str	Vienna, Burg, 26 Jan 1790	V/xix	II:5/xviii	Jan 1790 (V); vs (Leipzig, 1794); KMS 1789[β, γ, δ, ε]	8, 31, 32, 48, 55
620	620	Die Zauberflöte (Spl, 2, Schikaneder)	7 S, 2 A, 4 T, 5 B, SATB, 2 fl/pic, 2 ob, 2 cl/basset-hn, 2 bn, 2 hn, 2 tpt, 3 trbn, timp, glock, str	Vienna, auf der Wieden, 30 Sept 1791	V/xx	II:5/xix	mostly composed by July 1791 (V), ov. and march completed 28 Sept 1791 (V); excerpts, vs (Vienna, 1791–2): KMS 1791[a, α, b, β]	29, 31, 32, 33, 34, 53, 54, 55
621	621	La clemenza di Tito (os, 2, Metastasio, rev. C. Mazzolà)	4 S, T, B, SATB, 2 fl, 2 ob, 2 cl/basset-hn, 2 bn, 2 hn, 2 tpt, timp, str	Prague, National, 6 Sept 1791	V/xxi	II:5/xx	for Prague coronation of Leopold II; completed 5 Sept 1791 (V); plain recits not by Mozart; KMS 1791, [b, γ, δ, ε, ξ]	29, 32, 33, 53, 55

Music in: P. Anfossi: Il curioso indiscreto, Vienna, 1783; F. Bianchi: La villanella rapita, Vienna, 1785; Anfossi: Le gelosie fortunate, Vienna, 1788; D. Cimarosa: I due baroni, Vienna, 1789; U. Martín y Soler: Il burbero di buon cuore, Vienna, 1789

BALLET MUSIC

K	K6	Title	Scoring	Composition	MW	NMA	Remarks	
A10	299*b*	Les petits riens	2 fl, 2 ob, 2 cl, 2 bn, 2 hn, 2 tpt, timp, str	Paris, May–June 1778	XXIV, no.10a	II:6/ii, 13	perf. 11 June 1778, Paris, Opéra, after N. Piccinni: Le finte gemelle; 20 movts, ov. and 13 (of 20) by Mozart	16
300	300	[Gavotte]	2 ob, 2 bn, 2 hn, str	? Paris, early 1778		II:6/ii, 46	? discarded movt of Les petits riens	
367	367	[Ballet for Idomeneo]					see 'Operas'	
446	416*d*	[Pantomime]	str	Vienna, Feb 1783	XXIV, no.18	II:6/ii, 120	perf. Vienna, Hofburg, 3 March 1783; only 5 of at least 15 nos. extant	
A207	C27.06	[?for Ascanio in Alba]		? Milan, late 1771	—		9 nos. only extant, arr. pf; see Plath, D1964, 111–29	

Sketches: K299*c*, for a ballet of 27 nos., ? Paris, early 1778

Frag.: KA103/299*d*, La chasse (rondo), 2 fl, 2 ob, 2 bn, 2 hn, ? Paris, 1778, NMA, II:6/ii, 112

Doubtful: KA109/135*a*, Le gelosie del serraglio (for Lucio Silla), ? Milan, late 1772, autograph incipits for ballet of 32 nos., 6 from J. Starzer: Les cinque soltanes: see Senn, E 1961

DUETS AND ENSEMBLES FOR SOLO VOICES AND ORCHESTRA

K	K6	First words (author)	Voices	Accompaniment	Composition	MW	NMA	Remarks	
479	479	Dite almeno in che mancai (G. Bertati)	S, T, B, B	2 ob, 2 cl, 2 bn, 2 hn, str	Vienna, 5 Nov 1785 (C)	VI/ii, 70	II:7/iii, 101	for Bianchi: La villanella rapita, perf. Vienna, Burg, 28 Nov 1785	
480	480	Mandina amabile (Bertati)	S, T, B	2 fl, 2 ob, 2 cl, 2 bn, 2 hn, str	Vienna, 21 Nov 1785 (V)	VI/ii, 87	II:7/iii, 143	as K479 (Paris, 1789–90)	
489	489	Spiegarti non poss'io	S, T	2 ob, 2 bn, 2 hn, str	Vienna, 10 March 1786 (V)	V/xiii	II:5/xi, 376	for Idomeneo K366	25
540*b*	540*b*	Per queste tue manine (Da Ponte)	S, B	2 fl, 2 ob, 2 bn, 2 tpt, str	Vienna, 28 April 1788 (V)	V/xviii	II:5/xvii, 497	for Don Giovanni K527	
625	592*a*	Nun liebes Weibchen	S, B	fl, 2 ob, 2 bn, 2 hn, str	Vienna, Aug 1790	—	VI/2, 235	duet, for ? Schack: Der Stein der Weisen; ? partly orig; for other possible contribs. to opera see D.J. Buch, *COJ*, i (1997), 195–232	
615	615	Viviano felici (T. Grandi: *Le gelosie villane*)	S, A, T, B		Vienna, 20 April 1791 (V)	—	—	lost; known only from Mozart's catalogue; for perf. of G. Sarti: Le gelosie villane	

Frag.: K389/384*A*, Welch ängstliches Beben (Bretzner), T, T, fl, ob, bn, 2 hn, str, Vienna, April–May 1782, MW, XXIV, no.42, intended for Die Entführung aus dem Serail K384; K434/480b, Del gran regno delle amazzoni (*Petrosellini: Il regno delle amazoni*), T, B, B, 2 ob, 2 bn, 2 tpt, str, ? Vienna, end 1785, xxiv, no.44, II:7/iv, 154, 106 bars, inc. sketch, KMS 1785[b] = K626b/3

VOCAL ENSEMBLES WITH PIANO OR INSTRUMENTAL ENSEMBLE

K	K6	*First words (author)*	*Voices*	*Accompaniment*	*Composition*	*MW*	*NMA*	*Remarks*
A24*a*	43*a*	Ach, was müssen wir erfahren	S, S	?	? Vienna, Oct 1767	—	III:9, 51	? by L. Mozart
436	436	Ecco quel fiero istante (Metastasio: *Canzonette*)	S, S, B	3 basset-hn	? Vienna, ?1786	VI/ii, 65	III:9, 31	notturno; ? partly by G. von Jacquin; see Plath in Plath and others, D 1971–2
437	437	Mi lagnerò tacendo (Metastasio: *Siroe*)	S, S, B	2 cl, basset-hn	Vienna, 1786	VI/ii, 67	III:9, 35	as K436
438	438	Se lontan ben mio (Metastasio: *Strofe per musica*)	S, S, B	2 cl, basset-hn	Vienna, ?1786	XXIV, no.46	III:9, 29	as K436
439	439	Due pupille amabili	S, S, B	3 basset-hn	Vienna, ?1786	—	III:9, 26	as K436
346	439*a*	Luci care, luci belle	S, S, B	3 basset-hn	Vienna, ?1786	—	III:9, 42	as K436
441	441	Liebes Mandel, wo is's Bandel (?Mozart)	S, T, B	str	Vienna, ? early 1785 or ? 1786–7	VII/1, 25	III:9, 7	KMS 1786$^{\alpha, \beta}$
532	532	[Grazie agl'inganni tuoi] (Metastasio: *La libertà di Nice*)	S, T, B	fl, cl, 2 hn, 2 bn, b	? Vienna, 1787	VII/1, 73	III:9, 62	26 bars without words based on M. Kelly's duet 'Grazie agl'inganni tuoi'
549	549	Più non si trovano (Metastasio: *L'olimpiade*)	S, S, B	3 basset-hn	Vienna, 16 July 1788 (V)	VI/ii, 185	III:9, 44	authenticity of acc. doubtful

Frag.: KA5/571*a*, Caro mio Druck und Schluck (Mozart), S, T, T, B, ?pf, ? Vienna, 1789, MW, XXIV, no.50, NMA, III:9, 64

Spurious: K441*c*/C9.04, Liebes Mädchen, S, S, B, by M. Haydn, see Schmid, *ÖMz*, xxvi (1971), 72–9

ARIAS AND SCENES FOR VOICE AND ORCHESTRA

K	K6	*First words (author)*	*Accompaniment*	*Composition*	*MW*	*NMA*	*Remarks*
				for soprano			
23	23	Conservati fedele (Metastasio: *Artaserse*)	str	The Hague, Oct 1765	VI/i, 9; XXIV, no.54	II:7/i, 13	rev. Jan 1766
70	61*c*	A Berenice ... Sol nascente	2 ob, 2 hn, str	Salzburg, ?Dec 1766	VI/i, 23	II:7/i, 47	? licenza for Sarti: Vologeso, Salzburg, 28 Feb 1767, or for perf. March 1769
78	73*b*	Per pietà, bell'idol mio (Metastasio: *Artaserse*)	2 ob, 2 hn, str	*c*1765–6	VI/i, 49	II:7/i, 17	
A2	73A	Misero tu non sei (Metastasio: *Demetrio*)		Milan, 26 Jan 1770 (L)	—	—	lost; known only from letter, 26 Jan 1770
88	73*c*	Fra cento affanni (Metastasio: *Artaserse*)	2 ob, 2 hn, 2 tpt, str	Milan, Feb–March 1770	VI/i, 66	II:7/i, 65	9
79	73*d*	O temerario Arbace ... Per quel paterno amplesso (Metastasio: *Artaserse*)	2 ob, 2 bn, 2 hn, str	*c*1766	VI/i, 54	II:7/i, 23	
77	73*e*	Misero me ... Misero pargoletto (Metastasio: *Demofoonte*)	2 ob, 2 bn, 2 hn, str	Milan, March 1770	VI/i, 33	II:7/i, 83	9

K	K^6	*First words (author)*	*Accompaniment*	*Composition*	*MW*	*NMA*	*Remarks*	
82	73*o*	Se ardire, e speranza (Metastasio: *Demofoonte*)	2 fl, 2 hn, str	Rome, 25 April 1770 (D)	XXIV, no.48*a*	II:7/i, 103		
83	73*p*	Se tutti i mali miei (Metastasio: *Demofoonte*)	2 ob, 2 hn, str	Rome, April–May 1770	VI/i, 60	II:7/i, 115, 177	2 versions	
74*b*	74*b*	Non curo l'affetto (Metastasio: *Demofoonte*)	2 ob, 2 hn, str	Milan or Pavia, early 1771	—	II:7/i, 125	lacks authentic sources	
217	217	Voi avete un cor fedele (after Goldoni: *Le nozze di Dorina*)	2 ob, 2 hn, str	Salzburg, 26 Oct 1775 (D)	VI/i, 93	II:7/i, 147	? insertion for B. Galuppi: Le nozze di Dorina	15
272	272	Ah, lo previdi ... Ah, t'invola agl'occhi miei (Cigna-Santi: *Andromeda*)	2 ob, 2 hn, str	Salzburg, Aug 1777 (D)	VI/i, 119	II:7/ii, 23	for J. Dušek	
294	294	Alcandro lo confesso ... Non sò d'onde viene (Metastasio: *L'olimpiade*)	2 fl, 2 cl, 2 bn, 2 hn, str	Mannheim, 24 Feb 1778 (D)	VI/i, 134	II:7/ii, 41, 151	for A. Weber; 2 versions, KMS 1778[α]	15, 16
486*a*	295*a*	Basta vincesti ... Ah, non lasciarmi (Metastasio: *Didone abbandonata*)	2 fl, 2 bn, 2 hn, str	Mannheim, 27 Feb 1778 (D)	XXIV, no.61	II:7/ii, 77	for D. Wendling (i), inspired by an aria by Galuppi; see W. Plath, *Festschrift Walter Senn*, ed. E. Egg and E. Fässler (Munich, 1975), 174–8	
316	300*b*	Popoli di Tessaglia ... Io non chiedo (R. de' Calzabigi: *Alceste*)	ob, bn, 2 hn, str	Paris, July 1778; Munich, 8 Jan 1779 (D)	VI/i, 164	II:7/ii, 85	for A. Weber	
A3	315*b*	[Scena]	ob, 2 cl, 3 hn, pf, str	St Germain, Aug 1778	—	—	lost; for G.F. Tenducci; see Oldman, *ML*, xlii (1961), 44–52	
A11*a*	365*a*	Warum, o Liebe ... Zittre, töricht Herz (J.G. Dyck, after C. Gozzi: *Le due notti affannase*)		Munich, Nov 1780	—	—	partly lost; sung in Gozzi: Le due notti affannose, trans. F.A.C. Werther (Salzburg, 1 Dec 1780); see Edge, K1996	12
368	368	Ma che vi fece ... Sperai vicino (Metastasio: *Demofoonte*)	2 fl, 2 bn, 2 hn, str	Salzburg, 1779–80	VI/i, 183	II:7/ii, 107		
369	369	Misera! dove son ... Ah! non son io (Metastasio: *Ezio*)	2 fl, 2 hn, str	Munich, 8 March 1781 (D)	VI/i, 198	II:7/ii, 125	for Countess J. Paumgarten	19
374	374	A questo seno ... Or che il cielo (G. De Gamerra: *Sismano nel Mogol*)	2 ob, 2 hn, str	Vienna, April 1781 (L)	VI/i, 206	II:7/ii, 135	for F. Ceccarelli, perf. 8 April 1781	
119	382*b*	Der Liebe himmlisches Gefühl	? [2 ob, 2 hn, str]	?	XXIV, no.40	II:7/ii, 203	lacks authentic sources; acc. extant only in kbd red.	
383	383	Nehmt meinen Dank	fl, ob, bn, str	Vienna, 10 April 1782 (D)	VI/i, 217	II:7/iii, 3	? for A. Lange (née Weber)	
416	416	Mia speranza adorata ... Ah, non sai qual pena (G. Sertor: *Zemira*)	2 ob, 2 bn, 2 hn, str	Vienna, 8 Jan 1783 (D)	VI/ii, 2	II:7/iii, 11	for A. Lange, perf. 11 Jan and 23 March 1783	

K	K6	First words (author)	Accompaniment	Composition	MW	NMA	Remarks	
178	417*e*	Ah, spiegarti, oh Dio		Vienna, June 1783	XXIV, no.41	II:7/iii, 210	acc. extant only in kbd red., ? earlier version of K418	
418	418	Vorrei spiegarvi, oh Dio	2 ob, 2 bn, 2 hn, str	Vienna, 20 June 1783 (D)	VI/ii, 11	II:7/iii, 25	for A. Lange, insertion for Anfossi: Il curioso indiscreto, Vienna, Burg, 30 June 1783; KMS 1783[β]	23
419	419	No, che non sei capace	2 ob, 2 hn, 2 tpt, timp, str	Vienna, June 1783 (D)	VI/ii, 21	II:7/iii, 37	as K418; KMS 1783[d]	23
490	490	Non più, tutto ascoltai ... Non temer, amato bene	2 cl, 2 bn, 2 hn, vn solo, str	Vienna, 10 March 1786 (V)	V/xiii	II:5/xi, 192	see Idomeneo K366	25
505	505	Ch'io mi scordi di te ... Non temer, amato bene	2 cl, 2 bn, 2 hn, pf, str	Vienna, 26 Dec 1786 (D)	VI/ii, 100	II:7/iii, 175	for N. Storace; text from 1786 for Idomeneo K490	
528	528	Bella mia fiamma ... Resta, o cara (D.M. Scarcone: *Cerere placata*)	fl, 2 ob, 2 bn, 2 hn, str	Prague, 3 Nov 1787 (D, V)	VI/ii, 146	II:7/iv, 37	for J. Dušek	28
538	538	Ah se in ciel, benigne stelle (Metastasio: *L'eroe cinese*)	2 ob, 2 bn, 2 hn, str	Vienna, 4 March 1788 (D, V)	VI/ii, 161	II:7/iv, 57	for A. Lange; rev. of 1778 vocal part	
540*c*	540*c*	In quali eccessi ... Mi tradì (Da Ponte)	fl, 2 cl, bn, 2 hn, str	Vienna, 30 April 1788 (V)	V/xviii	II:5/xvii, 511	for Don Giovanni K527	
569	569	Ohne Zwang, aus eignem Triebe	2 ob, 2 bn, 2 hn, str	Vienna, Jan 1789 (V)	—	—	lost; Mozart's catalogue: 'Eine teutsche Aria'	24
577	577	Al desio di chi t'adora (?Da Ponte)	2 basset-hn, 2 bn, 2 hn, str	Vienna, July 1789 (V)	V/xvii	II:5/xvi, 602	rondò for A. Ferraresi del Bene, for Le nozze di Figaro K492; KMS 1789[α], see Page and Edge, K 1991	31
578	578	Alma grande e nobil core (G. Palomba)	2 ob, 2 bn, 2 hn, str	Vienna, Aug 1789 (V)	VI/ii, 187	II:7/iv, 91	for Cimarosa: I due baroni, Vienna, Burg, Sept 1789	31
579	579	Un moto di gioia (?Da Ponte)	fl, ob, bn, 2 hn, str	Vienna, Aug 1789	VII/1	II:5/xvi, 597	for Le nozze di Figaro K492	31
580	580	Schon lacht der holde Frühling	2 cl, 2 bn, 2 hn, str	Vienna, 17 Sept 1789 (V)	XXIV, no.48	II:7/iv, 168	for Ger. version of G. Paisiello: Il barbiere di Siviglia, not used; orch inc.	31
582	582	Chi sa qual sia (?Da Ponte)	2 cl, 2 bn, 2 hn, str	Vienna, Oct 1789 (V)	VI/ii, 195	II:7/iv, 105	for L. Villeneuve, for Martín y Soler: Il burbero di buon cuore, Vienna, Burg, 9 Nov 1789	31
583	583	Vado, ma dove? (?Da Ponte)	2 cl, 2 bn, 2 hn, str	Vienna, Oct 1789 (V)	VI/ii, 203	II:7/iv, 115	as K582	31
—	—	Quel destrier (Metastasio: *L'olimpiade*)		*c*1766			lost; Constanze owned MS, 1799; see letter, 13 Feb 1799	
—	—	Cara se le mie pene (?Metastasio: *Alessandro nell'Indie*)	2 hn, vn, va, b	? by 1772	—	II:7/i, 59	? = aria composed Olmütz, 1767	
—	—			Olmütz, Dec 1767 (L)	—	—	?lost, or = 'Cara se le mie pene'; see letter, 28 May 1778	

K	K⁶	*First words (author)*	*Accompaniment*	*Composition*	*MW*	*NMA*	*Remarks*	
—	—			Vienna, late sum. – aut. 1768	—	—	described in letters	
—	—			by Dec 1768	—	—	LC '15 Italian arias', incl. probably K21, 23, 78/73b, 79/73*d* and possibly 'Quel destrier'; 10 or 11 lost, not necessarily for S	
—	—	No caro fà corragio	str	Vienna, ? Aug 1790	—	—	acc. recit for aria by Cimarosa in P.A. Guglielmi: La Quaguera spiritosa, perf. Vienna, Burg, 13 Aug 1790; see A. Weinmann, 'Zur Mozart-Bibliographie', *Mozartgemenide Wien*, xlvii/June (1980), 3–7	

Frags.: K73*D*, Per quel paterno amplesso (Metastasio: *Artaserse*), 3 bars, *c*1766; K440/383*b*, In te spero (Metastasio: *Demofoonte*), 81 bars, v and b only, 1782 or ? later, MW, XXIV, no. 47

for alto

K	K⁶	*First words (author)*	*Accompaniment*	*Composition*	*MW*	*NMA*	*Remarks*	
255	255	Ombra felice ... Io ti lascio (De Gamera)	2 ob, 2 hn, str	Salzburg, Sept 1776 (D)	VI/i, 103	II:7/ii, 3	text from M. Mortellari: Arsace (Padua, 1775)	15

for tenor

K	K⁶	*First words (author)*	*Accompaniment*	*Composition*	*MW*	*NMA*	*Remarks*	
21	19*c*	Va dal furor portata (Metastasio: *Ezio*)	2 ob, 2 bn, 2 hn, str	London, 1765	VI/i, 1	II:7/i, 3, 163	2 versions, 1 rev. L. Mozart	
36	33*i*	Or che il dover ... Tali e cotanti sono	2 ob, 2 bn, 2 hn, 2 tpt, timp, str	Salzburg, Dec 1766	VI/i, 13	II:7/i, 33	licenza perf. anniversary of Archbishop Sigismund's consecration, 21 Dec 1766	
71	71	Ah più tremar non voglio (Metastasio: *Demofoonte*)	2 ob, 2 hn, str	? Italy 1770	XXIV, no.39	II:7/iv, 145	only 48 bars extant; ? continuation lost	
209	209	Si mostra la sorte	2 fl, 2 hn, str	Salzburg, 19 May 1775 (D)	VI/i, 83	II:7/i, 131		15
210	210	Con ossequio, con rispetto (Petrosellini: *L'astratto, ovvero Il giocator fortunato*)	2 ob, 2 hn, str	Salzburg, May 1775 (D)	VI/i, 87	II:7/i, 139		15
256	256	Clarice cara (Petrosellini: *L'astratto, ovvero Il giocator fortunato*)	2 ob, 2 hn, str	Salzburg, Sept 1776 (D)	VI/i, 113	II:7/ii, 15	? for Piccinni: L'astratto; KMS 1776[a]	
295	295	Se al labbro mio non credi	2 fl, 2 ob, 2 bn, 2 hn, str	Mannheim, 27 Feb 1778 (D)	VI/i, 148	II:7/ii, 59, 167	for A. Raaff; 2 versions; from J.A. Hasse: Artaserse, text attrib. A. Salvi	15
435	416*b*	Müsst'ich auch durch tausend Drachen	fl, ob, cl, 2 bn, 2 hn, 2 tpt, timp, str	? Vienna, early 1780s	XXIV, no.45	II:7/iv, 162	orch inc.; KMS 1783[α]	

K	K[6]	*First words (author)*	*Accompaniment*	*Composition*	*MW*	*NMA*	*Remarks*	
420	420	Per pietà, non ricercate	2 cl, 2 bn, 2 hn, str	Vienna, 21 June 1783 (D)	VI/ii, 31	II:7/iii, 51	for J.V. Adamberger, for Anfossi: Il curioso indiscreto, not perf.; KMS 1783[d]	23
431	425*b*	Misero! o sogno ... Aura che intorni spiri (Mazzolà: *L'isola capricciosa*)	2 fl, 2 bn, 2 hn, str	? Vienna, 1783	VI/ii, 39	II:7/iii, 81	for Adamberger	
540*a*	540*a*	Dalla sua pace (Da Ponte)	fl, 2 ob, 2 bn, 2 hn, str	Vienna, 24 April 1788 (D, C)	V/xviii	II:5/xviii, 489	for F. Morella, for Don Giovanni K527	
				for bass				
432	421*a*	Così dunque tradisci ... Aspri rimorsi atroci (Metastasio: *Temistocle*)	2 fl, 2 ob, 2 bn, 2 hn, str	? Vienna, *c*1782–3	VI/ii, 55	II:7/iii, 67		
512	512	Alcandro, lo confesso ... Non sò d'onde viene (Metastasio: *L'olimpiade*)	fl, 2 ob, 2 bn, 2 hn, str	Vienna, 19 March 1787 (D)	VI/ii, 120	II:7/iv, 3		
513	513	Mentre ti lascio (Angioli-Morbilli: *La disfatta di Dario*)	fl, 2 cl, 2 bn, 2 hn, str	Vienna, 23 March 1787 (D, C)	VI/ii, 133	II:7/iv, 19	for Jacquin	
539	539	Ich möchte wohl der Kaiser sein (J.W.L. Gleim)	pic, 2 ob, 2 bn, 2 hn, perc, str	Vienna, 5 March 1788 (D, C)	VI/ii, 177	II:7/iv, 79	Ger. warsong for F. Baumann, perf. Vienna, Leopoldstadt, 7 March 1788	30
541	541	Un bacio di mano (?Da Ponte)	fl, 2 ob, 2 bn, 2 hn, str	Vienna, May 1788 (C)	VI/ii, 180	II:7/iv, 83	for F. Albertarelli, for Anfossi: Le gelosie fortunate, Vienna, Burg, 2 June 1788	
584	584	Rivolgete a lui lo sguardo (Da Ponte)	2 ob, 2 bn, 2 tpt, timp, str	Vienna, Dec 1789 (C)	VI/ii, 209		for Così fan tutte K588; replaced by 'Non siate ritrosi'	
612	612	Per questa bella mano	fl, 2 ob, 2 bn, 2 hn, db solo, str	Vienna, 8 March 1791 (D, C)	VI/ii, 224	II:7/iv, 123	for F.X. Gerl and F. Pischelberger	33
A245	621a	Io ti lascio	str	? Prague, Sept 1791		II:7/iv, 139	? only vn parts by Mozart, rest by Jacquin; see U. Konrad, *MJb 1989–90*, 99–113	

Frags.: K209a, Un dente guasto, 16 bars, ? sum. 1772; K433/416*c*, Männer suchen stets zu naschen, ?mid-1780s, MW, XXIV, no.43, orch barely sketched

SONGS

with piano accompaniment unless otherwise stated

K	K6	Title	First words	Key	Author	Composition	MW	NMA	Remarks
53	47*e*	An die Freude	Freude, Königin der Weisen	F	J.P. Uz	Vienna, aut. 1768	VII/1, 2	III:8, 2	(Vienna, *c*1768)
147	125*g*		Wie unglücklich bin ich nit	F		Salzburg, ?1775–6	VII/1, 4	III:8, 4	masonic
148	125*h*	Lobegesang auf die feierliche Johannisloge	O heiliges Band der Freundschaft	D	L.F. Lenz	Salzburg, 1773	VII/1, 5	III:8, 4	masonic
307	284*d*	Ariette	Oiseaux, si tous les ans	C	A. Ferrand	Mannheim, wint. 1777–8	VII/1, 12	III:8, 6	for E.A. Wendling (ii)
308	295*b*	Ariette	Dans un bois solitaire	A♭	A.H. de la Motte	Mannheim, wint. 1777–8	VII/1, 14	III:8, 8	for E.A. Wendling (ii)
343	336*c*	[2 Ger. sacred songs]							see 'Short sacred works'
392	340*a*		Verdankt sei es dem Glanz	F	J.T. Hermes	Vienna, 1781–2	VII/1, 24	III:8, 15	
391	340*b*	[An die Einsamkeit]	Sei du mein Trost	B♭	Hermes	Vienna, 1781–2	VII/1, 23	III:8, 16	
390	340*c*	[An die Hoffnung]	Ich würd' auf meinem Pfad	d	Hermes	Vienna, 1781–2	VII/1, 22	III:8, 17	
349	367*a*	Die Zufriedenheit	Was frag ich viel	G	J.M. Miller	Munich, wint. 1780–81	VII/1, 18	III:8, 12	2 versions, one with mand acc.
351	367*b*		Komm, liebe Zither	C		Munich, wint. 1780–81	VII/1, 21	III:8, 14	mand acc.
A25	386*d*	[Gibraltar]	O Calpe!	D	J.N.C.M. Denis	Vienna, end 1782 (L)	—	III:8, 72	only pf part sketched
178	417*e*		Ah, spiegarti, o Dio						see 'Arias and Scenes ...' (soprano)
468	468	Lied zur Gesellenreise	Die ihr einem neuen Grade	B♭	J.F. von Ratschky	Vienna, 26 March 1785 (V)	VII/1, 34	III:8, 18	masonic; ? perf. Vienna, 16 April 1785; acc.: org in autograph, pf in Mozart's catalogue
472	472	Der Zauberer	Ihr Mädchen, flieht Damöten ja!	g	C.F. Weisse	Vienna, 7 May 1785 (V)	VII/1, 36	III:8, 20	(Vienna, 1788)
473	473	Die Zufriedenheit	Wie sanft, wie ruhig	B♭	Weisse	Vienna, 7 May 1785 (V)	VII/1, 38	III:8, 22	
474	474	Die betrogene Welt	Der reiche Tor	G	Weisse	Vienna, 7 May 1785 (V)	VII/1, 40	III:8, 24	(Vienna, 1788)
476	476	Das Veilchen	Ein Veilchen	G	J.W. von Goethe	Vienna, 8 June 1785 (V)	VII/1, 42	III:8, 26	(Vienna, 1789)
A11*a*	477*a*	Per la ricuperta salute die Ophelia			Da Ponte	Vienna, ? Sept 1785			lost, set by Mozart, Salieri and 'Cornetti', advertised in *Wienerblättchen*, 26 Sept 1785
483	483		Zerfliesset heut', geliebte Brüder	B♭	J.B. von Schloissnig	Vienna, end 1785	VII/1, 44	III:9, 20	masonic song, with male chorus
484	484		Ihr unsre neuen Leiter	G	Schloissnig	Vienna, end 1785	VII/1, 46	III:9, 22	masonic song, with male chorus

K	K^6	*Title*	*First words*	*Key*	*Author*	*Composition*	*MW*	*NMA*	*Remarks*
506	506	Lied der Freiheit	Wer unter eines Mädchens Hand	F	J.A. Blumauer	Vienna, ? end 1785	VII/1, 48	III:8, 28	(Vienna, 1786)
517	517	Die Alte	Zu meiner Zeit	e	F. von Hagedom	Vienna, 18 May 1787 (V)	VII/1, 50	III:8, 32	(Vienna, 1788)
518	518	Die Verschweigung	Sobald Damötas Chloen sieht	F	Weisse	Vienna, 20 May 1787 (V)	VII/1, 52	III:8, 34	inc. (Vienna, 1788), lost; later completions by ? J. André in autograph and by ? A.E. Müller; see U. Konrad, *MJb 1989–90*, 99–113
519	519	Das Lied der Trennung	Die Engel Gottes weinen	f	K.E.K. Schmidt	Vienna, 23 May 1787 (V)	VII/1, 54	III:8, 36	(Vienna, 1789)
520	520	Als Luise die Briefe	Erzeugt von heisser Phantasie	C	G. von Baumberg	Vienna, 26 May 1787 (D, C)	VII/1, 58	III:8, 40	
523	523	Abendempfindung	Abend ist's	F	?J.H. Campe	Vienna, 24 June 1787 (V)	VII/1, 60	III:8, 42	(Vienna, 1789)
524	524	An Chloe	Wenn die Lieb' aus deinen blauen	E♭	J.G. Jacobi	Vienna, 24 June 1787 (V)	VII/1, 64	III:8, 46	(Vienna, 1789)
529	529	Des kleinen Friedrichs Geburtstag	Es war einmal, ihr Leutchen	F	J.E.F. Schall	Prague, 6 Nov 1787 (V)	VII/1, 68	III:8, 50	(Vienna, 1788)
530	530	Das Traumbild	Wo bist du, Bild	E♭	L.H.C. Hölty	Prague, 6 Nov 1787	VII/1, 70	III:8, 52	circulated as work by Jacquin
531	531	Die kleine Spinnerin	Was spinnst du	C		Vienna, 11 Dec 1787 (V)	VII/1, 72	III:8, 54	(Vienna, 1787)
552	552	Beim Auszug in das Feld	Dem hohen Kaiser-Worte treu	A		Vienna, 11 Aug 1788 (V)	—	III:8, 56	(Vienna, 1788)
596	596	Sehnsucht nach dem Frühlinge	Komm, lieber Mai	F	C.A. Overbeck	Vienna, 14 Jan 1791 (V)	VII/1, 77	III:8, 58	(Vienna, 1791)
597	597	Im Frühlingsanfang	Erwacht zum neuen Leben	E♭	C.C. Sturm	Vienna, 14 Jan 1791 (V)	VII/1, 78	III:8, 59	(Vienna, 1791)
598	598	Das Kinderspiel	Wir Kinder	A	Overbeck	Vienna, 14 Jan 1791 (V)	VII/1, 80	III:8, 60	(Vienna, 1791)
619	619		Die ihr des unermesslichen Weltalls Schöpfer ehrt						cant., see 'Oratorios'
—	—	Zur Eröffnung der Meisterloge; Zum Schluss der Meisterarbeit	Des Todes Werk, der Faulniss Grauen; Vollbracht is die Arbeit der Meister		A. Veit von Schittlersber	Vienna, ? Aug 1785			masonic, ? perf. 12 Aug 1785; see Autexier, G 1984

K	K6	Title	First words	Key	Author	Composition	MW	NMA	Remarks
—	—	Bey Eröffnung der Tafelloge; Kettenlied; Lied in Nahmen der Armen	Legt für heut das werkzeug nieder; Wir singen, und schlingen zur Wette; Brüder! der blinde Greis am Stabe		G. Leon	Vienna, ? June–July 1790			masonic, ? perf. 6 July 1790; see Autexier, K 1992

Sketches, frags.: K—/441*a*, Ja! grüss dich Gott, 20 bars, ?Vienna, 1783; KA26/475a, Einsam bin ich, 8 bars; O Calpe! [Gibraltar] (J.N.C.M. Denis), Vienna, end 1782 (L), NMA, III:8, 27, only pf part sketched; K²—+A270–75, 277–83/C8.32–46, 15 Lieder (C.F. Gellert), ? by L. Mozart, see Plath and others, D1971–2; K—, Lustig sein die Schwobemedle, Salzburg, 1777–9

Doubtful: K52/46*c*, Daphne deine Rosenwangen, arr. by L. Mozart of Meiner liebsten schöne Wangen (Bastien und Bastienne K51/46*b*) with new text, MW, VII/1, 1, NMA, II:5/iii, 90

Spurious: K149/125*d*, Ich hab' es längst gesagt (Die grossmütige Gelassenheit) (L. Günther), MW, VII/1, 6; by L. Mozart; K150/125*e*, Was ich in Gedanken küsse (Geheime Liebe) (Günther), MW, VII/1, 7 by L. Mozart; K151/125*f*, Ich trachte nicht nach solchen Digen (Die Zufriedenheit) (F.R.L. von Canitz), MW, VII/1, 8, by L. Mozart; K152/210*a*, Ridente la calma (canzonetta), arr. ? by Mozart of aria by J. Mysliveček, see MW, VII/1, 9, M. Flothuis, *MJb 1971–2*, 241–3; K350/C.8.48, Wiegenlied, MW, VII/1, 20, by B. Flies

CANONS

K	K6	Work and type	Key	Composition	MW	NMA	Remarks, alternative texts
89*a*I	73*i*	canon 4 in 1	A	1772	—	III:10, 71	KMS 1772[a]
89	73*k*	Kyrie, 5 in 1	G	1772	III/i, 5	III:10, 3	
89*a*II	73*r*	1 Incipe Menalios, 3 in 1	F	1772	—	III:10, 73	
		2 Cantate Domino, 8 in 1	G				
		3 Confitebor, 2 in 1 (+ 1)	C				
		4 Thebana bella cantus, 6 in 2	B♭				
A109*d*	73*x*	14 canonic studies		1772			
229	382*a*	canon 3 in 1	c	? Vienna, *c*1782	VII/2, 2	III:10, 80	Sie ist dahin (L.H.C. Hölty)
230	382*b*	canon 2 in 1	c	? Vienna, *c*1782	VII/2, 4	III:10, 83	Selig, selig (Hölty)
231	382*c*	Leck mich im Arsch (Mozart), 6 in 1	B♭	? Vienna, *c*1782	VII/2, 5	III:10, 11	Lasst froh uns sein (C.G. Breitkopf)
233	382*d*	Leck mir den Arsch (Mozart), 3 in 1	B♭	? Vienna, *c*1782	VII/2, 11	III:10, 17	Nichts labt mich mehr (G.C. Härtel)
234	382*e*	Bei der Hitz' im Sommer ess ich (Mozart), 3 in 1	G	? Vienna, *c*1782	VII/2, 13	III:10, 20	Essen, trinken (Breitkopf)
347	382*f*	canon 6 in 1	D	? Vienna, *c*1782	VII/2, 15	III:10, 84	Wo der perlende Wein (Breitkopf): Lasst uns ziehn (L.V. Köchel)
348	382*g*	V'amo di core teneramente, 12 in 3	G	? Vienna, *c*1782	VII/2, 16	III:10, 24	
507	507	canon 3 in 1	F	Vienna, after 3 June 1786	VII/2, 18	III:10, 86	Heiterkeit und leichtes Blut (Härtel)
508	508	canon 3 in 1	F	Vienna, after 3 June 1786	VII/2, 18	III:10, 88	Auf das Wohl aller Freunde (Härtel)
—	508A	canon 3 in 1	C	Vienna, after 3 June 1786	—		
508*a*	508*a*, 1–2	2 canons 3 in 1	F	Vienna, after 3 June 1786	—	III:10, 89	
508*a*	508*a*, 3–8	6 canons 2 in 1	F	Vienna, after 3 June 1786	—	III:10, 90	

K	K⁶	*Work and type*	*Key*	*Composition*	*MW*	*NMA*	*Remarks, alternative texts*
232	509*a*	Lieber Freistädtler, lieber Gaulimauli (Mozart), 4 in 1	G	Vienna, after 4 July 1787	VII/2, 8; XXIV, no.52	III:10, 27	Wer nicht liebt Wein (Härtel)
283	515*b*	canon 4 in 2	F	Vienna, 24 April 1787 (D)	VII/2, 1	III:10, 96	Ach! zu kurz (Härtel)
553	553	Alleluia, 3 in 1	C	Vienna, 2 Sept 1788 (V)	VII/2, 19	III:10, 32	
554	554	Ave Maria, 4 in 1	F	Vienna, 2 Sept 1788 (V)	VII/2, 20	III:10, 34	
555	555	Lacrimoso son'io, 4 in 1	a	Vienna, 2 Sept 1788 (V)	VII/2, 21	III:10, 36	text earlier set by A. Caldara; Ach zum Jammer (Breitkopf)
556	556	Grechtelt's enk (Mozart), 4 in 1	G	Vienna, 2 Sept 1788 (V)	VII/2, 23	III:10, 38	Alles Fleisch (Breitkopf)
557	557	Nascoso e il mio sol, 4 in 1	f	Vienna, 2 Sept 1788 (V)	VII/2, 25	III:10, 40	text earlier set by Caldara
558	558	Gehn wir im Prater (Mozart), 4 in 1	B♭	Vienna, 2 Sept 1788 (V)	VII/2, 27	III:10, 43	Alles ist eitel hier (Breitkopf)
559	559	Difficile lectu mihi mars (Mozart), 3 in 1	F	Vienna, 2 Sept 1788 (V)	VII/2, 29	III:10, 47	Nimm, ist's gleich warm (Breitkopf)
560*a*	559*a*	O du eselhafter Peierl! (Mozart), 4 in 1	F	Vienna, 2 Sept 1788 (V)	VII/2, 36	III:10, 49, 55	versions K560, MW, VII/2, 31, in F or G with slightly different words; Gähnst du (Breitkopf)
561	561	Bona nox! bist a rechta Ox (Mozart), 4 in 1	A	Vienna, 2 Sept 1788 (V)	VII/2, 37	III:10, 62	Gute Nacht (Breitkopf)
562	562	Caro bell'idol mio, 3 in 1	A	Vienna, 2 Sept 1788 (V)	VII/2, 39	III:10, 65	text earlier set by Caldara; Ach süsses teures Leben (Breitkopf)
A191	562*c*	[? for 2 vn, va, b] 4 in 1	C	?Vienna	XXIV, no.51	III:10, 68	
—	—	canon 8 in 1	a	? Italy or Salzburg, 1770–71	—	—	see Zaslaw, D 1971–2
—	—	8 canons 2 in 1	F	Vienna, after 3 June 1786	—	III:10, 90	
—	—	canon 4 in 1	F	Vienna, ? sum. 1786	—	III:10, 97	

Spurious: KA109*d*/73*x*, 14 canonic studies, from G.B. Martini: *Storia della musica*; K562*a*, B♭, K562*b*, F, NMA, III:10, 98, by M. Haydn

SYMPHONIES, SYMPHONY MOVEMENTS

K	K⁶	*BH*	*Key*	*Movts*	*Scoring*	*Composition*	*MW*	*NMA*	*Remarks*	
16	16	1	E♭	3	2 ob, 2 hn, str	London, 1764– 5	VIII/i, 1	IV:11/i, 3		37
19	19	4	D	3	2 ob, 2 hn, str	London, 1765	VIII/i, 37	IV:11/i, 21		37
A223	19*a*	—	F	3	2 ob, 2 hn, str	London, 1765 – Paris, 1766	—	IV:11/i, 35		
22	22	5	B♭	3	2 ob, 2 hn, str	The Hague, Dec 1765	VIII/i, 47	IV:11/i, 49		37
43	43	6	F	4	2 ob/fl, 2 hn, str	? Salzburg–Vienna, 1767	VII/i, 56	IV:11/i, 79		
45	45	7	D	4	2 ob, 2 hn, 2 tpt, timp, str	Vienna, 16 Jan 1768 (D)	VIII/i, 69	IV:11/i, 95	adapted as ov. to La finta semplice	
A221	45*a*	—	G	3	2 ob, 2 hn, str	The Hague, 1766	—	IV:11/i, 115	'Lambach', rev. *c*1767	
A214	45*b*	—	B♭	4	2 ob, 2 hn, str	?Vienna, 1768	—	IV:11/i, 129	lacks authentic sources	

K	K6	BH	Key	Movts	Scoring	Composition	MW	NMA	Remarks	
48	48	8	D	4	2 ob, 2 hn, 2 tpt, timp, str	Vienna, 13 Dec 1768 (D)	VIII/i, 81	IV:11/i, 143		8, 37
73	73	9	C	4	2 ob/fl, 2 hn, 2 tpt, timp, str	Salzburg or Italy, 1769–70	VIII/ii, 97	IV:11/i, 163		
81	73*l*	44	D	3	2 ob, 2 hn, str	? Rome, April 1770	XXIV, no.4	IV:11/ii, 3	lacks authentic sources	9
97	73*m*	47	D	4	2 ob, 2 hn, 2 tpt, timp, str	? Rome, April 1770	XXIV, no.7	IV:11/ii, 15	lacks authentic sources	9
95	73*n*	45	D	4	2 fl, 2 tpt, str	? Rome, April 1770	XXIV, no.5	IV:11/ii, 33	lacks authentic sources	9
84	73*q*	11	D	3	2 ob, 2 hn, str	? Milan/Bologna, 1770	VIII/i, 21	IV:11/ii, 47	lacks authentic sources, also attrib. L. Mozart, C.D. von Dittersdorf and others; see J. LaRue, in Plath, L1971–2	9
74	74	10	G	3	2 ob, 2 hn, str	Milan, 1770	VIII/i, 110	IV:11/ii, 67		37
75	75	42	F	4	2 ob, 2 hn, str	Salzburg, 1771	XXIV, no.2	IV:11/ii, 83	lacks authentic sources	
110	75*b*	12	G	4	2 ob/fl, 2 bn, 2 hn, str	Salzburg, July 1771 (D)	VIII/i, 135	IV:11/ii, 97		11
120	111*a*	—	D	1	2 fl, 2 ob, 2 hn, 2 tpt, timp, str	? Milan, Oct–Nov 1771	XXIV, no.9	IV:11/ii, 115	finale, to form sym. with ov. to Ascanio in Alba K111	
96	111*b*	46	C	4	2 ob, 2 hn, 2 tpt, timp, str	? Milan, Oct–Nov 1771	XXIV, no.6	IV:11/ii, 133		
112	112	13	F	4	2 ob, 2 hn, str	Milan, 2 Nov 1771 (D)	VIII/i, 149	IV:11/ii, 151		11, 37
114	114	14	A	4	2 fl/ob, 2 hn, str	Salzburg, 30 Dec 1771 (D)	VIII/i, 161	IV:11/ii, 165		
124	124	15	G	4	2 ob, 2 hn, str	Salzburg, 21 Feb 1772 (D)	VIII/i, 175	IV:11/ii, 183		
128	128	16	C	3	2 ob, 2 hn, str	Salzburg, May 1772 (D)	VIII/i, 187	IV:11/iii, 1		
129	129	17	G	3	2 ob, 2 hn, str	Salzburg, May 1772 (D)	VIII/i, 199	IV:11/iii, 15		
130	130	18	F	4	2 fl, 4 hn, str	Salzburg, May 1772 (D)	VIII/i, 125	IV:11/iii, 31		
132	132	19	E♭	4	2 ob, 4 hn, str	Salzburg, July 1772 (D)	VIII/i, 233	IV:11/iii, 52	alternative slow movts: see L. Plath, *Mf*, xxvii (1974), 93–5	
133	133	20	D	4	fl, 2 ob, 2 hn, 2 tpt, str	Salzburg, July 1772 (D)	VIII/i, 252	IV:11/iii, 78		41
134	134	21	A	4	2 fl, 2 hn, str	Salzburg, Aug 1772 (D)	VIII/i, 271	IV:11/iii, 102		
161, 163	141*a*	50	D	3	2 fl, 2 ob, 2 hn, 2 tpt, timp, str	Salzburg, 1773–4	XXIV, no.10	IV:11/iii, 123	movts K161 from ov. to Il sogno di Scipione K126; K163 finale to form sym. with Il sogno di Scipione	
184	161*a*	26	E♭	3	2 fl, 2 ob, 2 bn, 2 hn, 2 tpt, str	Salzburg, 30 March 1773 (D)	VIII/ii, 58	IV:11/iv, 15		39, 40
199	161*b*	27	G	3	2 fl, 2 hn, str	Salzburg, ?10 April 1773 (D)	VIII/ii, 79	IV:11/iv, 37	date on MS possibly 16 April	
162	162	22	C	3	2 ob, 2 hn, 2 tpt, str	Salzburg, ?19 April 1773 (D)	VIII/ii, 1	IV:11/iv, 1	date on MS possibly 29 April	

K	K[6]	BH	Key	Movts	Scoring	Composition	MW	NMA	Remarks	
181	162*b*	23	D	3	2 ob, 2 hn, 2 tpt, str	Salzburg, 19 May 1773 (D)	VIII/ii, 13	IV:11/iv, 57		39
182	173*dA*	24	B♭	3	2 ob/fl, 2 hn, str	Salzburg, 3 Oct 1773 (D)	VIII/ii, 39	IV:11/iv, 75		
183	173*dB*	25	g	4	2 ob, 2 bn, 4 hn, str	Salzburg, 5 Oct 1773 (D)	VIII/ii, 39	IV:11/iv, 87		40
201	186*a*	29	A	4	2 ob, 2 hn, str	Salzburg, 6 April 1774 (D)	VIII/ii, 117	IV:11/v, 1		39, 40, 44
202	186*b*	30	D	4	2 ob, 2 hn, 2 tpt, str	Salzburg, 5 May 1774 (D)	VIII/ii, 141	IV:11/v, 26		
200	189*k*	28	C	4	2 ob, 2 hn, 2 tpt, timp, str	Salzburg, 17 [?12] Nov 1774 [?1773] (D)	VIII/ii, 95	IV:11/iv, 107		
121	207*a*	—	D	1	2 ob, 2 hn, str	Salzburg, end 1774 – early 1775	X, 42	IV:11/v, 44	finale, to form sym. with ov. to La finta giardiniera K196	
204	213*a*	—	D	4	2 ob, 2 bn, 2 hn, 2 tpt, str		—	IV:11/vii, 1	movts from Serenade K204/213*a*	
102	213*c*	—	D	1	2 ob/fl, 2 hn, 2 tpt, str	Salzburg, April and Aug 1775	XXIV, no.8	IV:11/v, 139	finale, to form sym. with versions of ov. and 1st aria of Il re pastore K208	
250	248*b*	—	D	4	2 ob, 2 bn, 2 hn, 2 tpt, timp, str		—	IV:11/vii, 31	movts from Serenade K250/248*b* with new timp part and other revs.	
297	300*a*	31	D	3	2 fl, 2 ob, 2 cl, 2 bn, 2 hn, 2 tpt, timp, str	Paris, June 1778	VIII/ii, 157	IV:11/v, 57	'Paris'; 2 slow movts, probable original in 1st edn (Paris, 1788), but see Tyson, D 1987; KMS 1778[a]	16, 42
318	318	32	G	1	2 fl, 2 ob, 2 bn, 4 hn, [2 tpt,] timp, str	Salzburg, 26 April 1779 (D)	VIII/ii, 197	IV:11/vi, 3	tpt part added 1782–3; possibly intended as ov. to Zaide K344/336*b*	17, 43
319	319	33	B♭	4	2 ob, 2 bn, 2 hn, str	Salzburg, 9 July 1779 (D)	VIII/ii, 213	IV:11/vi, 23	iii (minuet) added *c*1784–5; (Vienna, 1785) as op.7 no.2	17
320	320	—	D	3	2 ob, 2 bn, 2 hn, 2 tpt, timp, str		—	IV:11/viii, 89	movts from Serenade K320 with added timp	
338	338	34	C	3	2 ob, 2 bn, 2 hn, 2 tpt, timp, str	Salzburg, 29 Aug 1780 (D)	VIII/ii, 239	IV:11/vi, 59	frag. minuet (? originally complete) after 1st movt cancelled in autograph	17, 43
409	383*f*	—	C	1	2 fl, 2 ob, 2 bn, 2 hn, 2 tpt, timp, str	Vienna, May 1782	X, 48	IV:11/x, 3	mooted as intended for K338 although scoring differs	
385	385	35	D	4	2 fl, 2 ob, 2 cl, 2 bn, 2 hn, 2 tpt, timp, str	Vienna, July 1782	VIII/iii, 1	IV:11/vi, 113	'Haffner'; orig. intended as serenade, possibly with another minuet (lost) and March K408 no.2/385*a*; fls and cls later addns; (Vienna, 1785) as op.7 no.1	23, 42
425	425	36	C	4	2 ob, 2 bn, 2 hn, 2 tpt, timp, str	Linz, Oct–Nov 1783	VIII/iii, 37	IV:11/viii, 3	'Linz'; rev. Vienna, *c*1784–5; see Eisen, L 1988	23, 24
444	425*a*	37	G	1	2 ob, 2 hn, str	Vienna, late 1783 or 1784	VIII/iii, 81	—	introduction for M. Haydn: Sym. ST334/P16	

K	K⁶	BH	Key	Movts	Scoring	Composition	MW	NMA	Remarks	
504	504	38	D	3	2 fl, 2 ob, 2 bn, 2 hn, 2 tpt, timp, str	Vienna, 6 Dec 1786 (V)	VIII/iii, 97	IV:11/viii, 63	'Prague', last movt probably composed first; KMS 1786[b, γ]	27, 47
543	543	39	E♭	4	fl, 2 cl, 2 bn, 2 hn, 2 tpt, timp, str	Vienna, 26 June 1788 (V)	VIII/iii, 137	IV:11/ix, 1		49, 50
550	550	40	g	4	fl, 2 ob, [2 cl,] 2 bn, 2 hn, str	Vienna, 25 July 1788 (V)	VIII/iii, 181	IV:11/ix, 63	2 versions, 1st without cls also incl. rev. passage in slow movt; see Eisen, L 1997	49, 50
551	551	41	C	4	fl, 2 ob, 2 bn, 2 hn, 2 tpt, timp, str	Vienna, 10 Aug 1788 (V)	VIII/iii, 230	IV:11/ix, 187	'Jupiter'	49, 50, 51
A216	C11.03	54	B♭	4	2 ob, 2 hn, str		XXIV, no.63	—	K74g, lacks authentic sources; see G. Allroggen, *Mozart und Italien: Rome 1974* [*AnMc*, no. 18 (1978)], 237–45	

Sketches, frags.: KA100/383*g*, E♭, lost; K—/K³626*b*/34, KMS 1776[a], D, 64 bars, 'ouverture per un'opera buffa'; K—/K³467*a*/383*i*, KMS 1782[d], C, 2nd half of 1782, for sym. or ov.

Doubtful: KA222/19*b*, C; KA215/66*c*, D; KA217/66*d*, B♭; KA218/66*e*, B♭; K—/C11.07, G or D; K—/C11.08, F, lost, listed in Breitkopf & Härtel catalogue

Spurious: KA220/16*a*, 'Odense', see Zaslaw and Eisen, L 1985–6; K76/42*a*, F, NMA, IV:11/1, 63, see Eisen, L 1989; K18/A51, E♭, by C.F. Abel, see 'Arrangements'; K291/A52, C, by M. Haydn (ST287); K—/K³16*b*/C11.01, D by L. Mozart (C3); K17/C11.02, B♭, by L. Mozart (B♭6), MW, VIII/i, 13, see C. Eisen, *JAMS*, xxxix (1986), 615–32; K98/C11.04, MW, XXIV, F, no.56, attrib. 'Haydn' in *D-WEY*; K—/K³311*a*/C11.05, B♭, '2nd "Paris" symphony' (Paris, 1802–6); KA210/C11.06, D, by L. Mozart (D11); KA293/C11.09, G (Leipzig, 1841), by L. Mozart; K—/K³293*c*/C11.10, F, by I. Pleyel (B136); K—/C11.11, C, by A. Gyrowetz; K—/C11.12, F, by C. Ditters von Dittersdorf; KA294/C11.13, G by L. Mozart (G3); K—/C11.14, C, by A. Eberl, see S. Fischer, *MISM*, xxxi (1983), 21–6; K—, B♭, K—, D, both for 2 vn, b, *H-KE*, ? by L. Mozart

CASSATIONS, SERENADES, DIVERTIMENTOS, MISCELLANEOUS WORKS

K	K⁶	Title	Key, movts	Scoring	Composition	MW	NMA	Remarks	
32	32	Gallimathias musicum		hpd, 2 ob, 2 hn, 2 bn, str	The Hague, March 1766	XXIV, no.12	IV:12/i, 3		5
41*a*	41*a*	6 divertimentos		fl, hn, tpt, trbn, vn, va, vc	Salzburg, 1767	—	—	lost; in LC	
100	62*a*	Cassation	D, 8	2 ob/fl, 2 hn, 2 tpt, str	Salzburg, 1769	IX/i, 33	IV:12/i, 67	with March K62	8, 38
63	63	Cassation	G, 7	2 ob, 2 hn, str	Salzburg, 1769	IX/i, 1	IV:12/i, 25		8, 38
99	63*a*	Cassation	B♭, 7	2 ob, 2 hn, str	Salzburg, 1769	IX/i, 19	IV:12/i, 45		8, 38
113	113	Divertimento	E♭, 4	2 cl, 2 hn, (or 2 ob, ?2 cl, 2 eng hn, 2 bn, 2 hn), str	Milan, Nov 1771	IX/ii, 1	IV:12/ii, 3	'Concerto ò sia Divertimento'; rev. orch, early 1773, see Blazin, L 1992	11
136–8	125*a–c*	3 Divertimentos						see 'Chamber Music: String Quartets'	
131	131	Divertimento	D, 7	fl, ob, bn, 4 hn, str	Salzburg, June 1772	IX/ii, 15	IV:12/ii, 29		13
205	167*A*	Divertimento	D, 5	2 hn, bn, str (solo)	Salzburg, ?1773	IX/ii, 73	VII:18, 7	with March K290/167*AB*	13
185	167*a*	Serenade	D, 7	2 ob/fl, 2 hn, 2 tpt, vn solo, str	Vienna, July–Aug 1773	IX/i, 61	IV:12/ii, 76	with March K189/167*b*	13, 40

K	K6	Title	Key, movts	Scoring	Composition	MW	NMA	Remarks	
203	189*b*	Serenade	D, 8	2 ob/fl, bn, 2 hn, 2 tpt, vn solo, str	Salzburg, Aug 1774 (D)	IX/i, 97	IV:12/iii, 7	with March K237/189*c*	13, 40
204	213*a*	Serenade	D, 7	2 ob/fl, bn, 2 hn, 2 tpt, vn solo, str	Salzburg, 5 Aug 1775 (D)	IX/i, 133	IV:12/iii, 60	with March K215/213*b*; see also 'Symphonies'	14, 40
239	239	Serenata notturna	D, 3	2 vn, va, db (solo); str, timp	Salzburg, Jan 1776 (D)	IX/i, 177	IV:12/iii, 114		14
247	247	Divertimento	F, 6	2 hn, str (solo)	Salzburg, June 1776 (D)	IX/ii, 98	VII:18, 28	with March K248	14, 40
250	248*b*	Serenade	D, 8	2 ob/fl, 2 bn, 2 hn, 2 tpt, vn solo, str	Salzburg, June 1776 (D)	IX/i, 193	IV:12/iv, 8	'Haffner'; with March K249; see also 'Symphonies'	14, 40
251	251	Divertimento	D, 6	ob, 2 hn, str (solo)	Salzburg, July 1776 (D)	IX/ii, 121	VII:18, 67		
286	269*a*	Notturno	D, 3	4 groups, each 2 hn, str (solo)	Salzburg, Dec 1776 – Jan 1777	IX/i, 293	IV:12/v		
287	271*H*	Divertimento	B♭, 6	2 hn, str (solo)	Salzburg, June 1777	IX/ii, 168	VII:18, 103		40
320	320	Serenade	D, 7	2 fl/pic, 2 ob, 2 bn, 2 hn, post horn, 2 tpt, timp, str	Salzburg, 3 Aug 1779 (D)	IX/i, 325	IV:12/v	'Posthorn', with 2 marches, K335/320*a*; see also 'Symphonies', 'Concertos (wind instruments)'	17, 42
334	320*b*	Divertimento	D, 6	2 hn, str (solo)	Salzburg, 1779–80	IX/ii, 208	VI:18, 158	with March K445/320*c*	17
477	479*a*	Maurerische Trauermusik	c	2 ob, cl, 3 basset-hn, dbn, 2 hn, str	Vienna, 1785	X, 53	IV:11/x, 11	Dated July 1785 in (V); rev. version perf. Nov 1785; see Autexier, L 1984	26
522	522	Ein musikalischer Spass	F, 4	2 hn, str (solo)	Vienna, 14 June 1787 (C)	X, 58	VII:18, 223		
525	525	Eine kleine Nachtmusik	G, 4	2 vn, va, vc, b (solo)	Vienna, 10 Aug 1787 (V)	XIII, 181	IV:12/vi, 43	orig. 5 movts, 2nd lost	
—	—	Cassation	C		? Salzburg, 1769	—	—	lost; see letter, 18 Aug 1771	

Frags.: K288/246*c*, F, vn, va, b, 2 hn, 1775–7, NMA, VII:18, 260; K246*b*/320*B*, 2 hn, str, end 1772 – early 1773; KA108/522*a*, F, 2 hn, str, 1787, NMA, VIII:18, 266, ? related to K522; KA69/525*a*, C, 2 vn, va, vc, db, 1787, NMA, IV:2/vi, 66, ? related to K525

WIND ENSEMBLE

K	K^6	*Title*	*Key*	*Scoring*	*Composition*	*MW*	*NMA*	*Remarks*	
33*a*	33*a*	Solos		fl, [?bc]	Lausanne, Sept 1766	—	—	lost; in LC	
33*b*	33*b*	Piece		hn [+ ?]	? Salzburg, 1766	—	—	lost; mentioned in L. Mozart's letter, 16 Feb 1778	
41*b*	41*b*	Pieces		2 tpt/2 hn/2 basset-hn	Salzburg, 1767	—	—	lost; in LC	
186	159*b*	Divertimento	B♭	2 ob, 2 cl, 2 eng hn, 2 hn, 2 bn	Milan, March 1773	IX/ii, 57	VII:17/i		
166	159*d*	Divertimento	E♭	2 ob, 2 cl, 2 eng hn, 2 hn, 2 bn	Salzburg, 24 March 1773 (D)	IX/ii, 47	VII:17/i		
213	213	Divertimento	F	2 ob, 2 bn, 2 hn	Salzburg, July 1775 (D)	IX/ii, 83	VII:17/i		
240	240	Divertimento	B♭	2 ob, 2 bn, 2 hn	Salzburg, Jan 1776 (D)	IX/ii, 89	VII:17/i		
252	240*a*	Divertimento	E♭	2 ob, 2 bn, 2 hn	Salzburg, early 1776	IX/ii, 147	VII:17/i		
188	240*b*	Divertimento	C	2 fl, 5 tpt, timp	Salzburg, mid-1773	IX/ii, 69	VII:17/i		
253	253	Divertimento	F	2 ob, 2 bn, 2 hn	Salzburg, Aug 1776 (D)	IX/ii, 152	VII:17/i		
270	270	Divertimento	B♭	2 ob, 2 bn, 2 hn	Salzburg, Jan 1777 (D)	IX/ii, 159	VII:17/i		
361	370*a*	Serenade	B♭	2 ob, 2 cl, 2 basset-hn, 2 bn, 4 hn, db	Vienna, probably 1783–4	IX/i, 399	VII:17/ii, 141	see D.N. Leeson, *MJb 1997*, 181–223	24, 45
375	375	Serenade	E♭	[2 ob,] 2 cl, 2 bn, 2 hn	Vienna, Oct 1781	IX/i, 455	VII:17/ii, 3, 41	obs added in 2nd version, July 1782	45
388	384*a*	Serenade	c	2 ob, 2 cl, 2 bn, 2 hn	Vienna, ? July 1782 or late 1783	IX/i, 481	VII:17/ii, 97	arr. as str qnt, K406/516*b*	45
411	484*a*	Adagio	B♭	2 cl, 3 basset-hn	Vienna, 1782	X, 80	VII:17/ii, 223		
410	440*d*	Adagio	F	2 basset-hn, bn	Vienna, end 1782	X, 79	VIII:21, 120		
487	496*a*	12 Duos		2 hn [?basset-hn]	Vienna, 27 July 1786 (D)	XXIV, no.58	VIII:21, 49		

Frag., sketches: K384*B*, Andante, 2 ob, 2 cl, 2 bn, 2 hn, ?1782–3; K384*b*, March, 2 ob, 2 cl, 2 bn, 2 hn; K384*c*, Allegro, 2 ob, 2 cl, 2 bn, 2 hn, ?1782–3; KA95/484*b*, Allegro assai, B♭, 2 cl, 3 basset-hn, ?1786–7; KA93/484*c*, Adagio, F, cl, 3 basset-hn, ?1787–8 or later; K484*e*, Allegro, F, basset-hn, str inst; KA94/580*a*, Adagio, C, cl, 3 basset-hn (or eng hn, 2 hn/basset-hn, bn), 1780s

Doubtful: K289/271*g*, Divertimento, E♭, 2 ob, 2 bn, 2 hn, MW, IX/ii, 198, NMA, VII:7/i; KAC13.07, Partita, E♭, 2 ob, 2 cl, 2 bn, 2 hn, inc., *CZ-Pu*, see Leeson and Whitwell, N 1972; 4 partitas, 2 ob, 2 cl, 2 bn, 2 hn, F, B♭, E♭, B♭, incl. movts from KA17.04–05, arrs. of movts from K361/370*a*, movts in *CZ-Pu*; 5 pièces d'harmonie, 2 ob, 2 cl, 2 bn, 2 hn (Leipzig, 1802), incl. B♭, after K361/370*a*, E♭, KA226/C17.01, B♭, after K361/370*a*, B♭, KA227/C17.02, E♭, KA228/C17.03; 5 divertimentos, B♭ (2 basset-hn/cl, bn)/(3 basset-hn), MW, XXIV, no.62, NMA, VII:21, 67, 78, 89, 105, 114 (also 167), see Whewell, N 1962, 19, Flothuis, N 1973–4

Spurious: K187/C17.12, Divertimento, C, 2 fl, 5 tpt, timp, MW, IX/ii, 63, arr. by L. Mozart of dances by Starzer and Gluck; see also 'Arrangements', K626*b*, 28

MARCHES

K	K^6	Key	*Scoring*	*Composition*	*MW*	*NMA*	*Remarks*
41*c*	41*c*		2 ob, bn, 2 hn, 2 vn, b	Salzburg, 1767	—	—	lost; in LC
62	62	D	2 ob, 2 hn, 2 tpt, str	Salzburg, 1769	—	IV:12/i, 63	quoted in letter, 4 Aug 1770; used in Mitridate K87/74*a*; ? for Cassation K100/62*a*
290	167*AB*	D	2 hn, str	Salzburg, sum. 1772	X, 19	VII:18, 3	with Divertimento K205/167*A*
189	167*b*	D	2 fl, 2 hn, 2 tpt, 2 vn, b	Vienna, July–Aug 1773	X, 1	IV:12/ii, 70	with Serenade K185/167*a*
237	189*c*	D	2 ob, 2 bn, 2 hn, 2 tpt, 2 vn, b	Salzburg, Aug 1774	X, 10	IV:12/iii, 3	with Serenade K203/189*b*
215	213*b*	D	2 ob, 2 hn, 2 tpt, str	Salzburg, Aug 1775	X, 7	IV:12/iii, 55	with Serenade K204/213*a*
214	214	C	2 ob, 2 hn, 2 tpt, str	Salzburg, 20 Aug 1775 (D)	X, 4	IV:13/1/ii	
248	248	F	2 hn, str	Salzburg, June 1776 (D)	X, 13	VII:18, 23	with Divertimento K247
249	249	D	2 ob, 2 bn, 2 hn, 2 tpt, str	Salzburg, 20 July 1776 (D)	X, 16	IV:12/iv, 3	with Serenade K250/248*b*
335	320*a*	D	2 ob/fl, 2 hn, 2 tpt, str	Salzburg, Aug 1779	X, 22	IV:13/1/ii	2; with Serenade K320
445	320*c*	D	2 hn, str	Salzburg, sum. 1780	X, 114	VII:18, 155	with Divertimento K344/320*b*
408/1	383*e*	C	2 ob, 2 hn, 2 tpt, timp, str	Vienna, 1782	X, 28	IV:13/1/ii	
408/3	383*F*	C	2 fl, 2 bn, 2 hn, 2 tpt, timp, str	Vienna, 1782	X, 36	IV:13/1/ii	
408/2	385*a*	D	2 ob, 2 bn, 2 hn, 2 tpt, timp, str	Vienna, 1782	X, 32	IV:13/1/ii	
544	544	D	fl, hn, str	Vienna, June 1788 (V)	—	—	lost

DANCE MUSIC

K	K^6	*No.*	*Keys*	*Scoring*	*Composition*	*MW*	*NMA*	*Remarks*	
					Minuets (— without trio)*				
41*d*	41*d*			various	Salzburg, 1767	—	—	lost; in LC	
65*a*	61*b*	7	G, D, A, F, C, G, D	2 vn, b	Salzburg, 26 Jan 1769	XXIV, no.13	IV:13/1/i, 1		8
103	61*d*	19	C, G, D, F, C, A*, D, F, C, G, F, C, G, B♭, E♭, E*, A*, D, G*	2 ob/fl, 2 hn/tpt, 2 vn, b	Salzburg, spr.–sum. 1772	—	IV:13/1/i, 11, 78, 80	orig. 20; rearranged by Mozart as 19	8
104	61*e*							see 'Arrangements etc.'	8
—	61*g*II							see 'Arrangements etc.'	
122	73*t*	1	E♭*	2 ob, 2 hn, 2 vn, b	? Bologna, Aug 1770	XXIV, no.13*a*	IV:13/1/i, 10	see 'Arrangements etc.'	
164	130*a*	6	D, D, D, G, G, G	fl, ob, 2 hn/tpt, 2 vn, b	Salzburg, June 1772 (D)	XXIV, no.57	IV:13/1/i, 45		
176	176	16	C, G, E♭*, B♭*, F, D, A, C, G, B♭*, F, D, G, C, F, D	2 ob/fl, bn, 2 hn/tpt, 2 vn, b	Salzburg, Dec 1773 (D)		IV:13/1/i, 51	alternative versions of trios 1 and 2 also known	
363	363	3	D*, B♭*, D*	2 ob, 2 bn, 2 hn, 2 tpt, timp, 2 vn, b	? Vienna, *c*1782–3	XXIV, no.14	IV:13/1/ii		

K	K[6]	No.	Keys	Scoring	Composition	MW	NMA	Remarks
409	383*f*	1	C					see 'Symphonies'
461	448*a*	6	C, E♭, G, B♭, F, D*	2 ob/fl, 2 bn, 2 hn, 2 vn, b	Vienna, 1784	XI, 158	IV:13/1/ii	no.6 inc.
463	448*c*	2	F*, B♭*	2 ob, bn, 2 hn, 2 vn, b	Vienna, 1784	XI, 169	IV:13/1/ii	short minuets with contredanses
568	568	12	C, F, B♭, E♭, G, D, A, F, B♭, D, G, C	2 fl/pic, 2 ob/cl, 2 bn, 2 hn, 2 tpt, timp, 2 vn, b	Vienna, 24 Dec 1788 (V)	XI, 1	IV:13/1/ii	(Vienna, 1789)
585	585	12	D, F, B♭, E♭, G, C, A, F, B♭, E♭, G, D	2 fl/pic, 2 ob/cl, 2 bn, 2 hn, 2 tpt, timp, 2 vn, b	Vienna, Dec 1789 (V)	XI, 19	IV:13/1/ii	
599	599	6	C, G, E♭, B♭, F, D	2 fl/pic, 2 ob/cl, 2 bn, 2 hn, 2 tpt, timp, 2 vn, b	Vienna, 23 Jan 1791 (V)	XI, 37	IV:13/1/ii	transmitted with K601, 604
601	601	4	A, C, G, D	2 fl/pic, hurdy-gurdy, 2 ob/cl, 2 bn, 2 hn, 2 tpt, timp, 2 vn, b	Vienna, 5 Feb 1791 (V)	XI, 46	IV:13/1/ii	transmitted with K599, 604; composed with German Dances K602
604	604	2	B♭, E♭	2 fl, 2 cl, 2 bn, 2 tpt, timp, 2 vn, b	Vienna, 12 Feb 1791 (V)	XI, 53	IV:13/1/ii	transmitted with K599, 601; composed with German Dances K605

Doubtful: K61*g*[I], Menuet, NMA, IV:13/1/i, 40

Spurious: K105*f*/61*f*, 6 minuets, D, D, D, G, G, G, NMA, IV:13/1/i, by M. Haydn; K61*h*, 6 minuets, C, A*, D*, B♭, G, C, NMA, IV:13/1/i, 40, see Lindmayr-Brandl, L 1995; K315*a*, Minuet, by J.C. Bach

German dances, ländler

K	K[6]	No.	Keys	Scoring	Composition	MW	NMA	Remarks
509	509	6	D, G, E♭, F, A, C	2 fl/pic, 2 fl, 2 ob, 2 cl, 2 bn, 2 hn, 2 tpt, timp, 2 vn, b	Prague, 6 Feb 1787 (V)	XI, 56	IV:13/1/ii	
536	536	6	C, G, B♭, D, F, F	pic, 2 fl, 2 ob/cl, 2 bn, 2 hn/tpt, timp, 2 vn, b	Vienna, 27 Jan 1788 (V)	XI, 72	IV:13/1/ii	(Vienna, 1789)
567	567	6	B♭, E♭, G, D, A, C	pic, 2 fl, 2 ob/cl, 2 bn, 2 hn, 2 tpt, timp, 2 vn, b	Vienna, 6 Dec 1788 (V)	XI, 80	IV:13/1/ii	(Vienna, 1789)
571	571	6	D, A, C, G, B♭, D	2 fl/pic, 2 ob/cl, 2 bn, 2 hn/tpt, timp, perc, 2 vn, b	Vienna, 21 Feb 1789 (V)	XI, 92	IV:13/1/ii	
586	586	12	C, G, B♭, F, A, D, G, E♭, B♭, F, A, C	2 fl/pic, 2 ob/cl, 2 bn, 2 hn, 2 tpt, timp, perc, 2 vn, b	Vienna, Dec 1789 (V)	XI, 106	IV:13/1/ii	
600	600	6	C, F, B♭, E♭, G, D	pic, 2 fl, 2 ob/cl, 2 bn, 2 hn, 2 tpt, timp, 2 vn, b	Vienna, 29 Jan 1791 (V)	XI, 127	IV:13/1/ii	

K	K[6]	No.	Keys	Scoring	Composition	MW	NMA	Remarks
602	602	4	B♭, F, C, A	2 fl/pic, 2 ob/cl, 2 bn, 2 hn/tpt, timp, hurdy-gurdy, 2 vn, b	Vienna, 5 Feb 1791 (V)	XI, 139	IV:13/1/ii	with Minuets K601
605	605	3	D, G, C	2 fl/pic, 2 ob, 2 bn, 2 hn/tpt, 2 posthorns, timp, 5 sleighbells, 2 vn, b	Vienna, 12 Feb 1791 (V)	XI, 145	IV:13/1/ii	with Minuets K604; no.3, Die Schlittenfahrt, ? composed separately
606	606	6	B♭	2 vn, b [wind parts lost]	Vienna, 28 Feb 1791 (V)	XXIV, no.16	IV:13/1/ii	'Ländlerische', with Contredanse K607/605a
611	611	1	C	2 fl, 2 ob, 2 bn, 2 tpt, timp, hurdy-gurdy, 2 vn, b	Vienna, 6 March 1791 (V)	XI, 144	IV:13/1/ii	'Die Leyerer'; = K602, no.3

Contredanses

K	K[6]	No.	Keys	Scoring	Composition	MW	NMA	Remarks
123	73*g*	1	B♭	2 ob, 2 hn, 2 vn, b	Rome, 13–14 April 1770	XI, 152	IV:13/1/i, 7	
101	250*a*	4	F, G, D, F	2 ob/fl, bn, 2 hn, 2 vn, b	Salzburg, ? early 1776	IX/1, 57	IV:13/1/i, 67	'Serenade'
267	271*c*	4	G, E♭, A, D	2 ob/fl, bn, 2 hn, 2 vn, b	Salzburg, early 1777	XI, 154	IV:13/1/i, 71	
462	448*b*	6	C, E♭, B♭, D, B♭, F	2 ob, 2 hn, 2 vn, b	Vienna, Jan 1784	XI, 165	IV:13/1/ii	wind insts added later
463	448*c*	2	F, B♭	2 ob, bn, 2 hn, 2 vn, b	Vienna, Jan 1784	XI, 169	IV:13/1/ii	each preceded by a minuet
534	534	1	D	pic, 2 ob, 2 hn, side drum, 2 vn, b	Vienna, 14 Jan 1788 (V)	XXIV, no.27	IV:13/1/ii	Das Donnerwetter; extant only in pf red. and inc. orch parts
535	535	1	C	pic, 2 cl, bn, tpt, side drum, 2 vn, b	Vienna, 23 Jan 1788 (V)	XI, 184	IV:13/1/ii	La bataille (The Siege of Belgrade)
535*a*	535*a*	3	C, G, G		Vienna, ?early 1788			only pf version extant
565	565	2	B♭, D	2 ob, 2 hn, bn, 2 vn, b	Vienna, 30 Oct 1788 (V)	—	—	lost 24
587	587	1	C	fl, ob, bn, tpt, 2 vn, b	Vienna, Dec 1789 (V)	XI, 188	IV:13/1/ii	Der Sieg vom Helden Coburg
106	588*a*	3	D, A, B♭	2 ob, 2 bn, 2 hn, 2 vn, b	Vienna, Jan 1790	XXIV, no.15	IV:13/1/ii	with ov.
603	603	2	D, B♭	pic, 2 ob, 2 bn, 2 hn, 2 tpt, timp, 2 vn, b	Vienna, 5 Feb 1791 (V)	XI, 191	IV:13/1/ii	
607	605*a*	1	E♭	fl, ob, bn, 2 hn, 2 vn, b	Vienna, 28 Feb 1791 (V)	XXIV, no.17	IV:13/1/ii	Il trionfo delle dame; with German Dances K606
609	609	5	C, E♭, D, C, G	fl, side drum, 2 vn, b	Vienna, 1791	XI, 194	IV:13/1/ii	
610	610	1	G	2 fl, 2 hn, 2 vn, b	Vienna, 6 March 1791 (V)	XI, 200	IV:13/1/ii	Les filles malicieuses

K	K6	No.	Keys	Scoring	Composition	MW	NMA	Remarks
510	C13.02	9	D, D, D, B♭, D, D, F, B♭, C	2 pic, 2 ob/fl, 2 cl, 2 hn, 2 tpt, timp, 2 vn, b	? Prague, early 1787	XI, 173	IV:13/1/ii	probably not authentic

Frag.: KA107/535*b*, fl, ob, hn, bn, 2 vn, 1790–91 (? related to K603)

Doubtful: K—/269*b*, 12 contredanses, G, G, C, D, Salzburg, ? early 1776, nos.2, 12 = K101/250*a* nos.2, 3, see Eisen, B 1991, 269–70

CONCERTOS, CONCERTO MOVEMENTS

K	K6	BH	Key	Scoring	Composition	Cadenzas K624/626*a*	MW	NMA	Remarks	
					piano (all entitled 'Concerto')					
37, 39–41	37, 39–41	1–4							see 'Arrangements'	
107, 1–3	107, 1–3								see 'Arrangements'	
175	175	5	D	pf, 2 ob, 2 hn, 2 tpt, timp, str	Salzburg, Dec 1773 (D)	1–4	XVI/i, 131	V:15/i, 3	possibly for org; obs, 1st hn rev. 1777–8, see K. Hortschansky, *MJb 1989–90*; (Vienna, 1785) as op.7; see K382	13, 20, 40
238	238	6	B♭	pf, 2 ob/fl, 2 hn, str	Salzburg, Jan 1776 (D)	5–7	XVI/i, 165	V:15/i, 89		14
242	242	7	F	3 pf, 2 ob, 2 hn, str	Salzburg, Feb 1776 (D)	—	XVI/i, 197	V:15/i, 155	'Lodron'; also version for 2 pf	14
246	246	8	C	pf, 2 ob, 2 hn, str	Salzburg, April 1776 (D)	8–14	XVI/i, 289	V:15/ii, 3	'Lützow'	14
271	271	9	E♭	pf, 2 ob, 2 hn, str	Salzburg, Jan 1777 (D)	15–22	XVI/ii, 1	V:15/ii, 65	'Jeunehomme'	14, 41
365	316*a*	10	E♭	2 pf, 2 ob, 2 bn, 2 hn, str	Salzburg, ? late 1780	23–4	XVI/ii, 53	V:15/ii, 145	for dating see Konrad, M 1990	17
382	382	—	D	pf, fl, 2 ob, 2 hn, 2 tpt, timp, str	Vienna, March 1782	25–6	XVI/iv, 359	V:15/i, 67	new finale for K175	20
414	385*p*	12	A	pf, 2 ob, 2 hn, str	Vienna, 1782	27–36	XVI/ii, 133	V:15/iii, 3	(Vienna, 1785) as op.4 no.1; KMS 1782^{d} = K—/385*o*	23, 24
386	386	—	A	pf, 2 ob, 2 hn, str	Vienna, 19 Oct 1782 (D)	—	—	V:15/viii, 173	? intended as finale for K414/385*p*; inc.	
413	387*a*	11	F	pf, 2 ob, 2 bn, 2 hn, str	Vienna, 1782–3	37–8	XVI/ii, 101	V:15/iii, 67	(Vienna, 1785) as op.4 no.2	23, 24
415	387*b*	13	C	pf, 2 ob, 2 bn, 2 hn, 2 tpt, timp, str	Vienna, 1782–3	39–41	XVI/ii, 163	V:15/iii, 127	(Vienna, 1785) as op.4 no.3; cancelled slow movt, 16 bars, in autograph	20, 23, 24
449	449	14	E♭	pf, 2 ob, 2 hn, str	Vienna, 9 Feb 1784 (D, V)	42	XVI/ii, 205	V:15/iv, 3	probably begun 1782–3; for Barbara Ployer	50

	K^6	BH	Key	Scoring	Composition	Cadenzas K624/626a	MW	NMA	Remarks	
450	450	15	B♭	pf, fl, 2 ob, 2 bn, 2 hn, str	Vienna, 15 March 1784 (V)	43–5	XVI/ii, 241	V:15/iv, 67		24, 50, 51
451	451	16	D	pf, fl, 2 ob, 2 bn, 2 hn, 2 tpt, timp, str	Vienna, 22 March 1784 (V)	46–7	XVI/ii, 285	V:15/iv, 137	(Paris, *c*1785); ornamentation of ii, K624/626*a*II, *M*	24
453	453	17	G	pf, fl, 2 ob, 2 bn, 2 hn, str	Vienna, 12 April 1784 (V)	48–51	XVI/iii, 22	V:15/v, 3	for Barbara Ployer; (Speyer, 1789) as op.9	
456	456	18	B♭	pf, fl, 2 ob, 2 bn, 2 hn, str	Vienna, 30 Sept 1784 (V)	52–7	XVI/iii, 55	V:15/v, 71	'Paradies'	51
459	459	19	F	pf, fl, 2 ob, 2 bn, 2 hn, str	Vienna, 11 Dec 1784 (V)	58–60	XVI/iii, 119	V:15/v, 151		45
466	466	20	d	pf, fl, 2 ob, 2 bn, 2 hn, 2 tpt, timp, str	Vienna, 10 Feb 1785 (V)	—	XVI/iii, 181	V:15/vi, 3	38-bar false start, last movt, in autograph	24, 45
467	467	21	C	pf, fl, 2 ob, 2 bn, 2 hn, 2 tpt, timp, str	Vienna, 9 March 1785 (V)	—	XVI/iii, 237	V:15/vi, 93		51
482	482	22	E♭	pf, fl, 2 cl, 2 bn, 2 hn, 2 tpt, timp, str	Vienna, 16 Dec 1785 (V)	—	XVI/iv, 1	V:15/vi, 177		
488	488	23	A	pf, fl, 2 cl, 2 bn, 2 hn, str	Vienna, 2 March 1786 (V)	61	XVI/iv, 67	V:15/vii, 3		
491	491	24	c	pf, fl, 2 ob, 2 cl, 2 bn, 2 hn, 2 tpt, timp, str	Vienna, 24 March 1786 (V)	—	XVI/iv, 121	V:15/vii, 85		28, 50
503	503	25	C	pf, fl, 2 ob, 2 bn, 2 hn, 2 tpt, timp, str	Vienna, 4 Dec 1786 (V)	—	XVI/iv, 185	V:15/vii, 256	KMS 1786[b]	49, 52
537	537	26	D	pf, fl, 2 ob, 2 bn, 2 hn, 2 tpt, timp, str	Vienna, 24 Feb 1788 (V)	—	XVI/iv, 253	V:15/viii, 3	'Coronation', pf part inc.; KMS 1787[c]	28
595	595	27	B♭	pf, fl, 2 ob, 2 bn, 2 hn, str	Vienna, 5 Jan 1791 (V)	62–4	XVI/iv, 309	V:15/viii, 93	possibly begun 1788; (Vienna, 1791) as op.17	33

Frags.: KA65/452*c*, slow movt, C, ?1784–6, NMA, V:15/vii, 188; KA59/459A, slow movt, C, ?1784; KA58/488*a*, slow movt, D, 1785–6, NMA, V:15/vii, 191; KA63/488*b*, ?rondo, A, ?1785–6, NMA, V:15/vii, 192; KA64/488*c*, ?rondo, A, 1785–6, NMA, V:15/vii, 193; K—/488*d*, rondo, A, 1785–6, NMA, V:15/vii, 194; KA62/491*a*, slow movt, E♭, ?1786, NMA, V:15/vii, 195; KA60/502*a*, first movt, C, 1784–5, NMA, V:15/vii, 196; KA57/537*a*, first movt, D, 1785–6, NMA, V:15/vii, 197; KA61/537*b*, slow movt, d, ? late 1786, NMA, V:15/vii, 198

Frag., vn, pf solos: KA56/315*f*, D, Mannheim, 1778, MW, XXIV, no.21a, NMA, V:14/ii, 136

Wolfgang Amadeus Mozart: portrait (early 1770) probably by Saverio dalla Rosa, formerly attributed to Felice Cignaroli (private collection); the music he is playing is presumed to be by Mozart himself (this is its only source) and is entered in the Köchel catalogue (3/1937) as K72*a*

Mozart wearing the insignia of the Golden Spur; anonymous portrait (1777) in the Civico Museo Bibliographico Musicale, Bologna

Wolfgang Amadeus Mozart: unfinished portrait (probably 1789) by Joseph Lange in the Mozart Museum, Salzburg

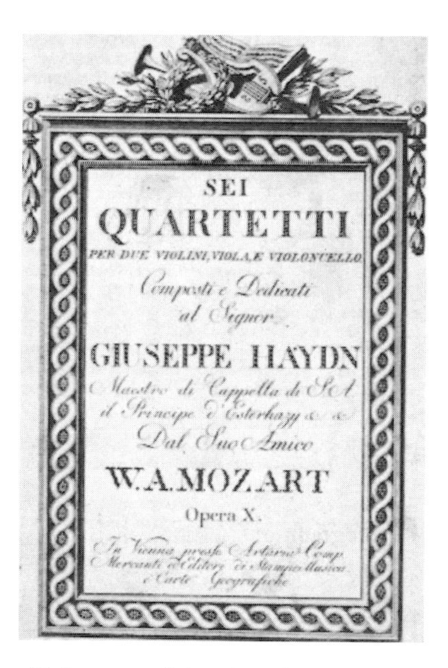

SEI
QUARTETTI
PER DUE VIOLINI, VIOLA, E VIOLONCELLO
Composti e Dedicati
al Signor
GIUSEPPE HAYDN
Maestro di Cappella di S.A.
il Principe d'Esterhazy &c. &c.
Dal Suo Amico
W.A.MOZART
Opera X.
In Vienna presso Artaria Comp.
Mercanti ed Editori di Stampe Musica
e Carte Geografiche

Title-page of the six string quartets, K387, 421/417*b*, 428/421*b*, 458, 464 and 465, dedicated to Joseph Haydn

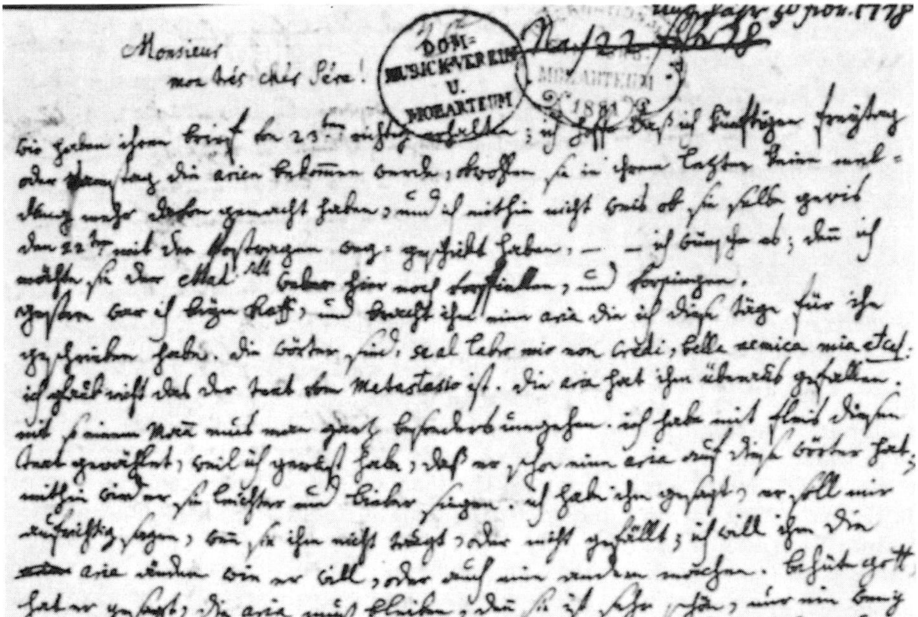

Monsieur
mon très cher Père!

Autograph letter from Mozart to his father (dated '30' February 1778) written from Mannheim; it refers to the aria 'Se al labbro' K295 that he composed for Anton Raaff (Mozart Museum, Salzburg)

First page of the autograph MS of Mozart's 'Posthorn' Serenade K320, composed 1779 (*D-Bsb*)

1784.
den 9.ten Hornung.

1. Ein klavierkonzert. Begleitung. 2 violini, viola e Basso. (2 oboe, 2 corni ad libitum)

den 15.ten März.

2. Ein klavier konzert. Begleitung. 2 violini, 2 viole, 1 flauto, 2 oboe, 2 fagotti, 2 corni e Basso.

den 22.ten

3. Ein klavierkonzert. Begleitung. 2 violini, 2 viole, 1 flauto, 2 oboe, 2 fagotti, 2 corni, 2 clarini, timpany e Basso.

den 30.ten

4. Ein klavier Quintett. Begleitung. 1 oboe, 1 clarinetto, 1 corno, et 1 fagotto.

den 12.ten aprill.

5. Ein klavier konzert. Begleitung. 2 violini, 2 viole, 1 flauto, 2 oboe, 2 fagotti, 2 corni, e Basso.

Allegro.

oboe e fagotti — Violini — All.o

All.o

Largo.

All.o

The first page of Mozart's 'Verzeichnüss aller meiner Werke', begun in February 1784 (*GB-Lbl Loan 42/1*)

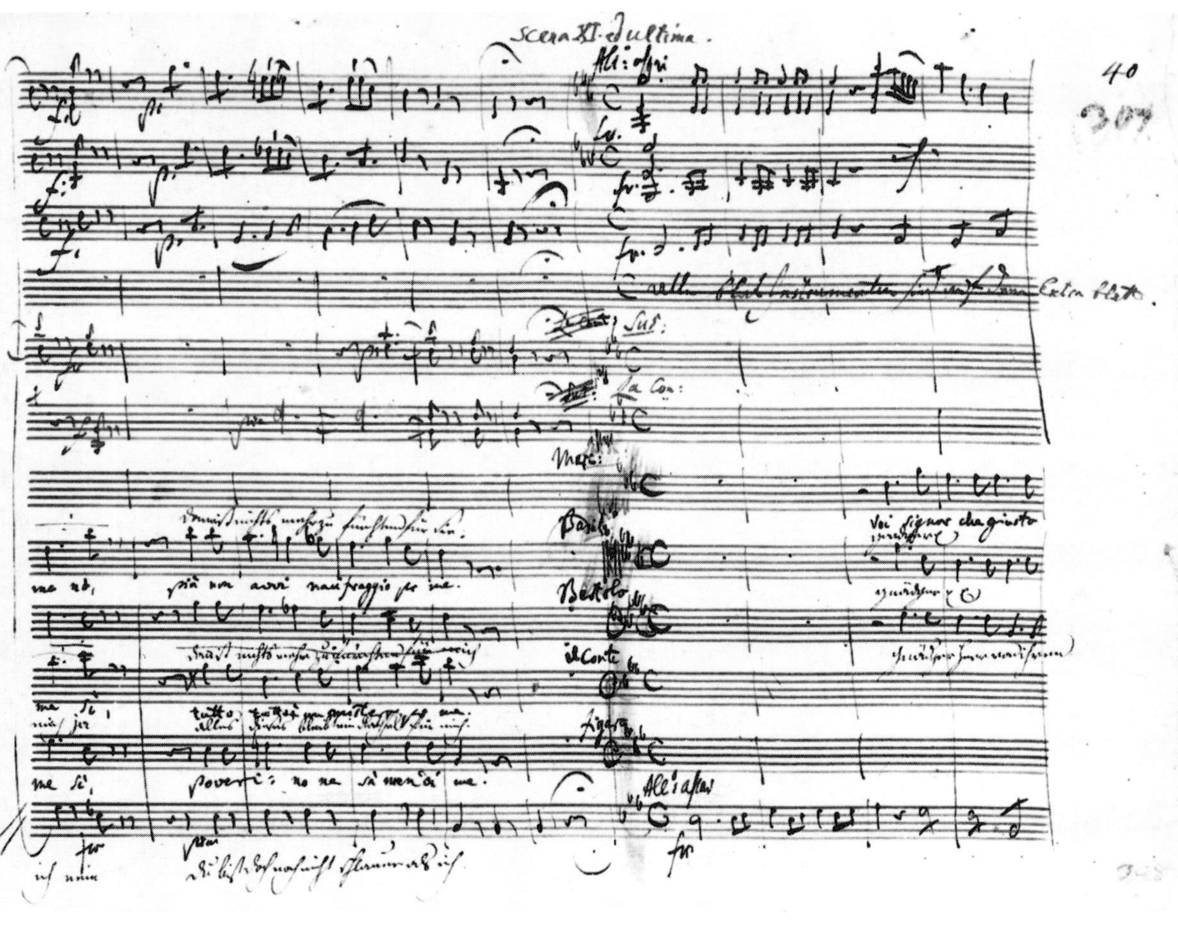

A page from the autograph MS of Mozart's 'Le nozze di Figaro' K492, 1786 (Act 2 finale, beginning of the last scene) (*B-Bsb*)

Autograph MS from Mozart's String Quartet in D K575, 1789, end of the first movement (*GB-Lbl* Add.37765, f.4v)

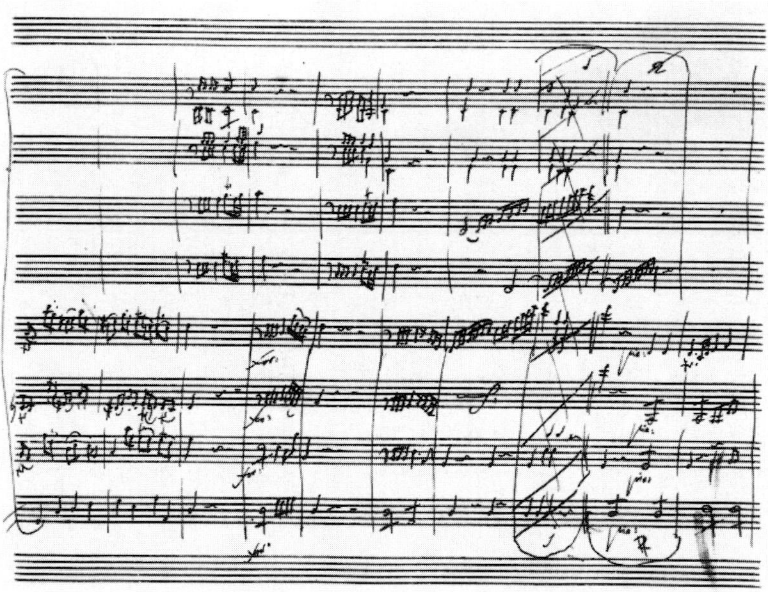

Autograph MS of the symphony K133, showing the end of the recapitulation

Mit gnädigster Erlaubniß

Wird Heute Freytags den 15ten October 1790.

im grosen Stadt-Schauspielhause

Herr Kapellmeister Mozart

ein groses

musikalisches Konzert

zu seinem Vortheil geben.

Erster Theil.

Eine neue grose Simphonie von Herrn Mozart.

Eine Arie, gesungen von Madame Schick.

Ein Concert auf dem Forte-piano, gespielt von Herrn Kapellmeister Mozart von seiner eigenen Komposition.

Eine Arie, gesungen von Herrn Cecarelli.

Zweyter Theil.

Ein Concert von Herrn Kapellmeister Mozart von seiner eigenen Komposition.

Ein Duett, gesungen von Madame Schick und Herrn Cecarelli.

Eine Phantasie aus dem Stegreife von Herrn Mozart.

Eine Symphonie.

Die Person zahlt in den Logen und Parquet 2 fl. 45 kr.

Auf der Gallerie 24 kr.

Billets sind bey Herrn Mozart, wohnhaft in der Kalbsbechergasse Nro. 167. vom Donnerstag Nachmittags und Freytags Frühe bey Herrn Cassirer Scheidweiler und an der Casse zu haben.

Der Anfang ist um Eilf Uhr Vormittags.

Handbill for the concert Mozart gave in Frankfurt (15 October 1790) during the festivities on the coronation of Leopold II; the programme included the piano concertos K537 and 459

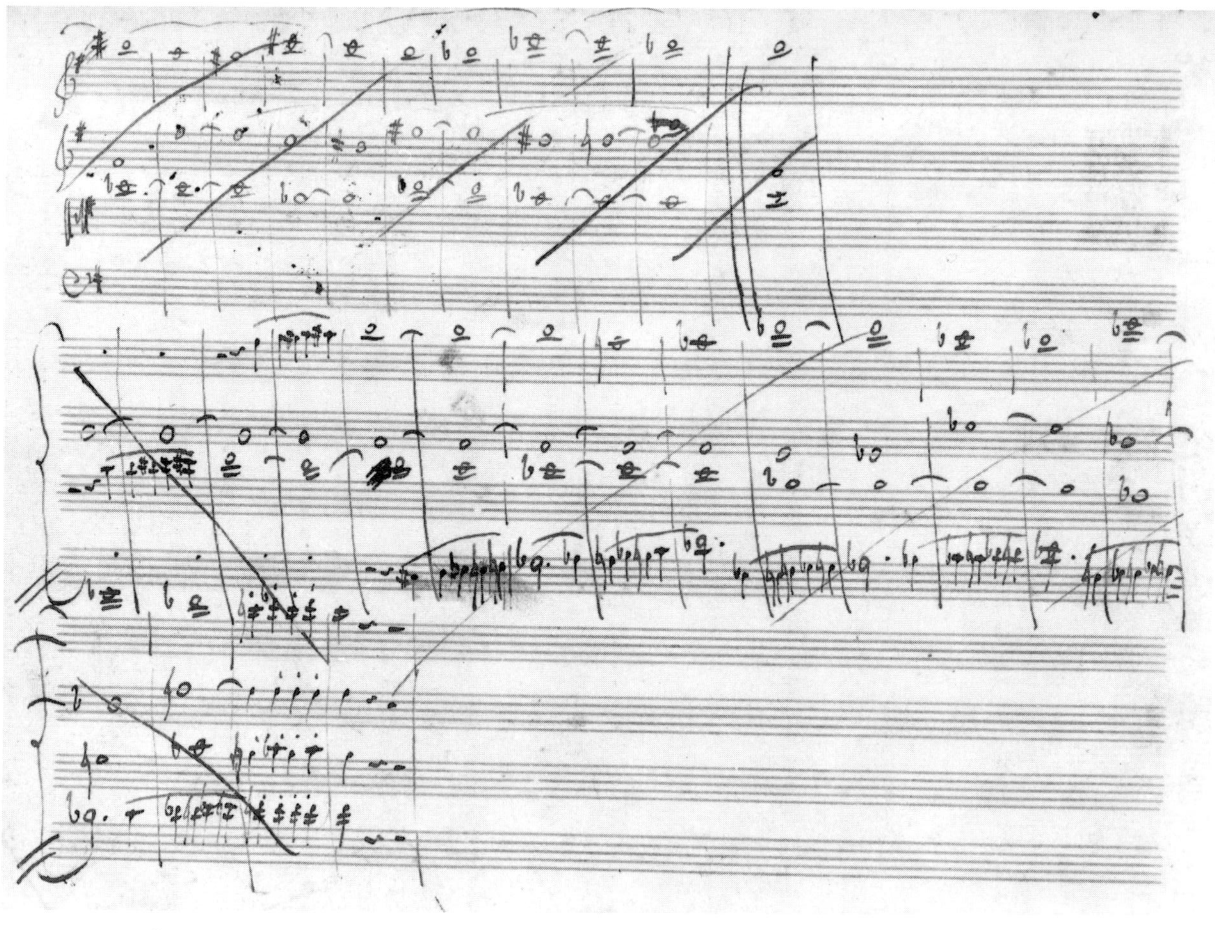

Autograph MS of the String Quartet K387, corrections and revisions to an enharmonic passage (Seiffert plate 2 Fol. 12v)

strings

K	K⁶	*Title*	*Key*	*Solo*	*Accompaniment*	*Composition*	*MW*	*NMA*	*Remarks*	
190	186*E*	Concertone	C	2 vn	solo ob, vc; 2 ob, 2 hn, 2 tpt, str	Salzburg, 31 May 1774 (D)	XII/i, 167	V:14/ii, 3		13
207	207	Concerto	B♭	vn	2 ob, 2 hn, str	Salzburg, 1773	XII/i, 1	V:14/i, 3	date on autograph 14 April 1775, but originally '1773'	14
211	211	Concerto	D	vn	2 ob, 2 hn, str	Salzburg, 14 June 1775 (D)	XII/i, 27	V:14/i, 55		14
216	216	Concerto	G	vn	2 ob, 2 hn, str	Salzburg, 12 Sept 1775 (D)	XII/i, 49	V:14/i, 95		14, 41
218	218	Concerto	D	vn	2 ob, 2 hn, str	Salzburg, Oct 1775 (D)	XII/i, 83	V:14/i, 151		14, 41
219	219	Concerto	A	vn	2 ob, 2 hn, str	Salzburg, 20 Dec 1775 (D)	XII/i, 113	V:14/i, 205		14, 41
261	261	Adagio	E	vn	2 fl, 2 hn, str	Salzburg, 1776	XII/i, 145	V:14/i, 267	for K219	
269	261*a*	Rondo	B♭	vn	2 ob, 2 hn, str	Salzburg, 1776	XII/i, 150	V:14/i, 275	? for K207	
364	320*d*	Sinfonia concertante	E♭	vn, va	2 ob, 2 hn, str	Salzburg, 1779–80	XII/i, 211	V:14/ii, 57	for dating see Konrad, M 1990; KMS 1779$^{\beta1-2}$	17, 43
373	373	Rondo	C	vn	2 ob, 2 hn, str	Vienna, 2 April 1781 (D)	XII/i, 159	V:14/i, 293		
470	470	Andante	A	vn	2 ob, 2 hn, str	Vienna, 1 April 1785 (V)	—	—	lost; ? for concerto	24

Frag.: KA104/320*e*, Sinfonia concertante, vn, va, vc solos, A, 1779, NMA, V:14/ii, 153

Doubtful: K—/206*a*, F, vc, lost; K—/K²271*a*/271*i*, D, vn, NMA, X:29/i, 81, see King, M1978, 31–2, and Mahling, *MJb 1978–9*, 252–68

Spurious: K268/C14.04, E♭, vn, MW, XXIV/xix, ? by J.F. Eck, see W. Lebermann, *MF*, xxxi (1978), 45–6; KA294*a*/C14.05, 'Adelaide Concerto', D, vn, by its 'editor', H. Casadesus (Mainz, 1930)

wind

K	K⁶	*Title*	*Key*	*Solo*	*Accompaniment*	*Composition*	*MW*	*NMA*	*Remarks*	
—	47*c*	Concerto		tpt		Vienna, *c*Nov 1768	—	—	lost, perf. Vienna, Waisenhauskirche, 7 Dec 1768; see L. Mozart's letter, 12 Nov 1768	8
191	186*e*	Concerto	B♭	bn	2 ob, 2 hn, str	Salzburg, 4 June 1774 (D)	XII/ii, 1	V:14/iii, 133		
271*k*	271*k*	Concerto		ob		Salzburg, 1777	—	—	mentioned in letter, 14 Feb 1778; ?lost, or K314/285*d*, KMS 1779$^{\alpha}$	
313	285*c*	Concerto	G	fl	2 ob, 2 hn, str	Mannheim, early 1778	XII/ii, 73	V:14/iii, 3	lacks authentic sources	
314	285*d*	Concerto	C	ob/fl	2 ob, 2 hn, str	? Mannheim, early 1778	XII/ii, 104	V:14/iii, 53, 97	? = K271*k*, also arr. fl, D, authenticity uncertain; sk for ob 1777$^{\alpha}$	
315	285*e*	Andante	C	fl	2 ob, 2 hn, str	? Mannheim, early 1780	XII/ii, 129	V:14/iii, 89		

K	K[6]	Title	Key	Solo	Accompaniment	Composition	MW	NMA	Remarks	
A9	297*B*	Sinfonia concertante		fl, ob, bn, hn		? Paris, April 1778	—	—	lost; ? partly transmitted by KA9/C14.01, E♭, ob, cl, bn, hn solos, acc. 2 ob, 2 hn, str, MW, XXIV, no.7*a*, NMA, X:29/i, 3; see Levin, M 1988	16
299	297*c*	Concerto	C	fl, hp	2 ob, 2 hn, str	Paris, April 1778	XII/ii, 21	V:14/vi, 3		
320	320	Sinfonia concertante	G	2 fl, 2 ob, 2 bn					iii and iv of Serenade K320; see letter of 29 March 1783	
412	386*b*	Concerto	D	hn	2 ob, 2 hn, str	Vienna, 1791	XII/ii, 135	V:14/v, 89	ii inc.; version in *RUS-SPit* (K514) is 1792 completion by F.X. Süssmayr	
417	417	Concerto	E♭	hn	2 ob, 2 hn, str	Vienna, 27 May 1783	XII/ii, 149	V:14/v, 3		16
447	447	Concerto	E♭	hn	2 cl, 2 bn, str	Vienna, ?1787	XII/ii, 167	V:14/v, 29		
495	495	Concerto	E♭	hn	2 ob, 2 hn, str	Vienna, 26 June 1786 (V)	XII/ii, 187	V:14/v, 57		
622	622	Concerto	A	cl	2 fl, 2 bn, 2 hn, str	Vienna, Oct 1791	XII/ii, 207	V:14/v, 3	orig. solo part, with range to written *c'*, lost; draft of i, G, basset-hn = K584*b*/621*b*	33, 55

Frag. movts.: K—/370*b*, E♭, hn, ?1781, NMA, V:14/v, 105; K371, E♭, hn, ?1781, MW, XXIV, no.21, NMA, V:14/iii; K293/416*f*, F, ob, Paris or Mannheim, 1778, MW, XXIV, no.20, NMA, V:14/iii, 167; K—/494*a*, E, hn, ?1783–7, NMA, V:14/v, 121; K584*b*/621*b*, G, basset-hn, 1790–91, NMA, V:14/iv, 165, = 1st movt of K622

Doubtful: KA230/196*d*, F, bn; ? others for bn, lost

Spurious: K—/C14.03, B♭, bn

CHAMBER

K	K[6]	Title	Key	Scoring	Composition	MW	NMA	Remarks	
					Strings and wind				
285	285	Quartet	D	fl, vn, va, vc	Mannheim, 25 Dec 1777 (D)	XIV, 307	VIII:20/2, 3		15
285*a*	285*a*	Quartet	G	fl, vn, va, vc	Mannheim, Jan–Feb 1778		VIII:20/2, 25		
298	298	Quartet	A	fl, vn, va, vc	Vienna, 1786–7	XIV, 310	VIII:20/2, 51		
370	368*b*	Quartet	F	ob, vn, va, vc	Munich, early 1781	XIV, 327	VIII:20/2, 65		19
407	386*c*	Quintet	E♭	hn, vn, 2 va, vc	Vienna, end 1782	XIII, 41	VIII:19/2, 1		
581	581	Quintet	A	cl, 2 vn, va, vc	Vienna, 29 Sept 1789 (V)	XIII, 112	VIII:19/2, 15		33, 51

Frags.: KA91/516*c*, B♭, and K516*d*, E♭, cl, 2 vn, va, vc; KA90/580*b*, F, cl, basset-hn, vn, va, vc; KA88/581*a*, A

Doubtful: K292/196*c*, Duo, B♭, bn, vc, MW, X, 75, NMA, VIII:21, 7, (Leipzig, 1805); KA171/285*b*, Quartet, C, fl, vn, va, vc, KMS 1781, (Speyer, 1788) as op.14, ii arr. from Serenade K361/370*a*, see R. Lustig, *MJb 1997*, 157–79

String quintets: 2 violins, 2 violas, cello

K	K^6	Key	Composition	MW	NMA	Remarks	
174	174	B♭	Salzburg, Dec 1773	XIII, 1	VIII:19/1, 3		13
515	515	C	Vienna, 19 April 1787 (V)	XIII, 54	VIII:19/1, 27	(Vienna, 1789)	45
516	516	g	Vienna, 16 May 1787 (V)	XIII, 85	VIII:19/1, 63	(Vienna, 1790)	43, 45, 52
406	516*b*	c	Vienna, 1788	XIII, 23	VIII:19/1, 91	arr. from Serenade K388/384*a*	
593	593	D	Vienna, Dec 1790 (V)	XIII, 132	VIII:19/1, 113		31, 33, 52
614	614	E♭	Vienna, 12 April 1791 (V)	XIII, 156	VIII:19/1, 143		31, 33, 52, 53

Frags.: KA80/514*a*, B♭, ?1787; KA87/515*a*, F, ?1791; KA79/515*c*, a, ?1791; KA86/516*a*, g, ? May 1787, related to K516; KA81/613*a*, E♭, late 1784 – 1785; KA83/592*b*, D, ?1788; KA2/613*b*, ?1786–7/?1789

Doubtful: 3 preludes, see 'Arrangements'

Spurious: K46, MW, XXIV, no.22, arr. of movts from Serenade K361/370*a*

String quartets

K	K^6	Title	Key	Composition	MW	NMA	Remarks	
80	73*f*	Quartet	G	Lodi, 15 March 1770 (D)	XIV, 1; XXIV, no.55	VIII:20/1/i, 3	iv added Vienna, late 1773, or Salzburg, early 1774	9
136	125*a*	Divertimento	D	Salzburg, early 1772	XIV, 278	IV:12/vi, 3		
137	125*b*	Divertimento	B♭	Salzburg, early 1772	XIV, 287	IV:12/vi, 19		
138	125*c*	Divertimento	F	Salzburg, early 1772	XIV, 294	IV:12/vi, 30		
155	134*a*	[Quartet]	D	Bolzano, Verona, Oct–Nov 1772	XIV, 8	VIII:20/1/i, 17		
156	134*b*	Quartet	G	Milan, end 1772	XIV, 15	VIII:20/1/i, 31		
157	157	Quartet	C	Milan, end 1772–early 1773	XIV, 21	VIII:20/1/i, 41		
158	158	Quartet	F	Milan, end 1772–early 1773	XIV, 29	VIII:20/1/i, 57		
159	159	Quartet	B♭	Milan, early 1773	XIV, 36	VIII:20/1/i, 69		
160	159*a*	Quartet	E♭	Milan, early 1773	XIV, 45	VIII:20/1/i, 85		
168	168	Quartet	F	Vienna, Aug 1773	XIV, 52	VIII:20/1/i, 99		13, 40
169	169	Quartet	A	Vienna, Aug 1773 (D)	XIV, 60	VIII:20/1/i, 113		13
170	170	Quartet	C	Vienna, Aug 1773 (D)	XIV, 69	VIII:20/1/i, 129		13
171	171	Quartet	E♭	Vienna, Aug 1773	XIV, 77	VIII:20/1/i, 145		13
172	172	Quartet	B♭	Vienna, ? Sept 1773	XIV, 86	VIII:20/1/i, 159		13
173	173	Quartet	d	Vienna, [Sept] 1773 (D)	XIV, 96	VIII:20/1/i, 175		13, 40

K	K6	Title	Key	Composition	MW	NMA	Remarks	
387	387	Quartet	G	Vienna, 31 Dec 1782 (D)	XIV, 106	VIII:20/1/ii, 3	(Vienna, 1785) as op.10 no.1; sketch in autograph	23, 45
421	417*b*	Quartet	d	Vienna, June 1783	XIV, 124	VIII:20/1/ii, 33	(Vienna, 1785) as op.10 no.2	23
428	421*b*	Quartet	E♭	Vienna, June–July 1783	XIV, 137	VIII:20/1/ii, 85	(Vienna, 1785) as op.10 no.4	44
458	458	Quartet	B♭	Vienna, 9 Nov 1784 (V)	XIV, 152	VIII:20/1/ii, 57	'Hunt' (Vienna, 1785) as op.10 no.3	44, 51
464	464	Quartet	A	Vienna, 10 Jan 1785 (V)	XIV, 168	VIII:20/1/ii, 111	(Vienna, 1785) as op.10 no.5	44, 51
465	465	Quartet	C	Vienna, 14 Jan 1785 (V)	XIV, 186	VIII:20/1/ii, 145	'Dissonance' (Vienna, 1785) as op.10 no.6	45
499	499	Quartet	D	Vienna, 19 Aug 1786 (V)	XIV, 206	VIII:20/1/iii, 3	'Hoffmeister' (Vienna, 1786)	25
546	546	Adagio and Fugue	c	Vienna, 26 June 1788 (V)	XIV, 301	IV:11/x, 47	? for str orch; fugue arr. from K426	45
575	575	Quartet	D	Vienna, June 1789 (V)	XIV, 226	VIII:20/1/iii, 37	'Prussian'	31
589	589	Quartet	B♭	Vienna, May 1790 (V)	XIV, 242	VIII:20/1/iii, 65	'Prussian'	31, 52
590	590	Quartet	F	Vienna, June 1790 (V)	XIV, 258	VIII:20/1/iii, 93	'Prussian'	52

Frags.: K168*a*, F, early 1775; KA77/405*a*, C, *c*1790; KA76/417*c*, after 1786; K417*d*, e, *c*1789; g, with K453*b*, ?1783; KA75/458*a*, B♭, KA71/458*b*, B♭, both *c*1790, NMA, VIII:20/1/iii, ? related to K589; KA72/464*a*, A, *c*1784, related to K464; KA47/587*a*, g, *c*1789; KA68/589*a*, B♭, *c*1783, NMA, VIII:20/1/iii, 148; KA73/589*b*, F, *c*1790, NMA, VIII:20/1/iii, 149, ? related to K590; K—, E, 1782–3

Doubtful: 6 preludes, see 'Arrangements'

Spurious: B♭, C, A, E♭, KA210–13/C20.01–4, ed. H. Wollheim (Mainz, 1932), by J. Schuster; see Finscher, N1966

String sonatas, duos, trios

K	K6	Title	Key	Scoring	Composition	MW	NMA	Remarks	
33*b*	33*b*	Solos		vc, b	Donaueschingen, Oct 1766	—	—	lost; in LC (incipit ? = 2nd pt of that in K6 for K196*d*)	
41*g*	41*g*	Nachtmusik		2 vn, b	? Salzburg, 1767	—	—	lost; see N. Mozart's letter, 8 Feb 1800	
46*d*	46*d*	Sonata	C	vn, b	Vienna, 1 Sept 1768 (D)	—	VIII:21, 3		
46*e*	46*e*	Sonata	F	vn, b	Vienna, 1 Sept 1768 (D)	—	VIII:21, 5		
266	271*f*	Trio	B♭	2 vn, b	Salzburg, early 1777	XXIV, no.23	VIII:21, 61		
404*a*	404*a*	4 preludes		vn, va, vc	Vienna, 1782	—	—	doubtful; for fugues by J.S. and W.F. Bach; see 'Arrangements'	
423	423	Duo	G	vn, va	? Salzburg or Vienna, 1783	XV, 1	VIII:21, 15		
424	424	Duo	B♭	vn, va	? Salzburg or Vienna, 1783	XV, 9	VIII:21, 33		
563	563	Trio	E♭	vn, va, vc	Vienna, 27 Sept 1788 (V)	XV, 19	VIII:21, 121	'Ein Divertimento ... di sei pezzi'	51
—	—			b viol, b		—	—	lost; in LC (incipit ? as K33*b*)	
—	—	6 trios		2 vn, vc	before 1768	—	—	lost; in LC	

Frags.: K443/404*b*, Fugue, G, completed by M. Stadler; KA66/562*e*, G, vn, va, vc, *c*1790–91; K—, Trio, 2 vn, vc, Vienna, 1785–6 or later

Keyboard and two or more instruments

K	K⁶	Title	Key	Scoring	Composition	MW	NMA	Remarks	
10–15	10–15	6 sonatas		hpd, vn [, vc]				see 'Keyboard and violin'	
254	254	Divertimento	B♭	pf, vn, vc	Salzburg, Aug 1776 (D)	XVII/2, 2	VIII:22/2, 56	(Paris, *c*1782) as op.3	
452	452	Quintet	E♭	pf, ob, cl, bn, hn	Vienna, 30 March 1784 (V)	XVII/1, 2; XXIV, no.59	VIII:22/1, 107	sk KA54/452*a*; KMS 1783[δ]	45, 52
478	478	Quartet	g	pf, vn, va, vc	Vienna, 16 Oct 1785 (D)	XVII/1, 32	VIII:22/1, 1	(Vienna, 1785–6)	25
493	493	Quartet	E♭	pf, vn, va, vc	Vienna, 3 June 1786 (V)	XVII/1, 62	VIII:22/1, 53	(Vienna, 1787) as op.13; KMS 1786[d]	25
496	496	Trio	G	pf, vn, vc	Vienna, 8 July 1786 (V)	XVII/2, 46	VIII:22/2, 78	(Vienna, 1786); sk	25, 44
498	498	Trio	E♭	pf, cl, va	Vienna, 5 Aug 1786 (V)	XVII/2, 68	VIII:22/2, 104	(Vienna, 1788) as op.14	45
502	502	Trio	B♭	pf, vn, vc	Vienna, 18 Nov 1786 (V)	XVII/2, 86	VIII:22/2, 129	(Vienna, 1788) as op.15 no.1	44, 52
542	542	Trio	E	pf, vn, vc	Vienna, 22 June 1788 (V)	XVII/2, 110	VIII:22/2, 160	(Vienna, 1788) as op.15 no.2	25, 44
548	548	Trio	C	pf, vn, vc	Vienna, 14 July 1788 (V)	XVII/2, 132	VIII:22/2, 188	(Vienna, 1788) as op.15 no.3	25, 44
564	564	Trio	G	pf, vn, vc	Vienna, 27 Oct 1788 (V)	XVII/2, 150	VIII:22/2, 212	(London, 1789)	27
617	617	Adagio and Rondo	c	armonica, fl, ob, va, vc	Vienna, 23 May 1791 (V)	X, 85	VIII:22/1, 146		

Frags.: K442, d, pf, vn, vc, Vienna, ?1783–90, MW, XVII/2, 20, inc., completed by M. Stadler, ? 3 separate movts, d, G, D, associated fortuitously; KA54/452*a*, B♭, kbd, ob, cl, basset-hn, bn, ?1785, ? related to K452; KA53/493*a*, E♭, pf, vn, va, vc, *c*1786, ? related to K493; KA52/495*a*, G, pf, vn, vc, *c*1786–7, NMA, VIII:22/2, 271, ? related to K496; KA51/501*a*, B♭, pf, vn, vc, 1784–5; KA92/616*a*, C, armonica, fl, ob, va, vc, ?1791, NMA, VIII:22/1, 168, related to K617

Keyboard and violin

K	K⁶	Title	Key	Composition	MW	NMA	Remarks	
6–7	6–7	2 Sonatas	C, D	Salzburg, Paris, 1762–4	XVIII/i, 2, 12	VIII:23/i, 2, 12	(Paris, 1764) as op.1	3, 4
8–9	8–9	2 Sonatas	B♭, G	Paris, 1763–4	XVIII/i, 20, 26	VIII:23/i, 20, 26	(Paris, 1764) as op.2	4, 36
10–15	10–15	6 Sonatas	B♭, G, A, F, C, B♭	London, 1764	XVIII/i, 34, 42, 47, 54, 62, 72	VIII:22/2, 2, 12, 18, 26, 36, 48	(London, 1765) as op.3; vc ad lib	
26–31	26–31	6 Sonatas	E♭, G, C, D, F, B♭	The Hague, Feb 1766	XVIII/i, 78, 84, 90, 96, 100, 106	VII:23/i, 34, 40, 45, 50, 54, 59	(The Hague and Amsterdam, 1766) as op.4	5
301	293*a*	Sonata	G	Mannheim, early 1778	XVIII/ii, 18	VIII:23/i, 66	(Paris, 1778) as op.1 no.1	42
302	293*b*	Sonata	E♭	Mannheim, early 1778	XVIII/ii, 32	VIII:23/i, 78	(Paris, 1778) as op.1 no.2	42
303	293*c*	Sonata	C	Mannheim, early 1778	XVIII/ii, 44	VIII:23/i, 88	(Paris, 1778) as op.1 no.3	42
305	293*d*	Sonata	A	Mannheim, early 1778	XVIII/ii, 64	VIII:23/i, 107	(Paris, 1778) as op.1 no.5	42
296	296	Sonata	C	Mannheim, 11 March 1778 (D)	XVIII/ii, 2	VIII:23/i, 139	(Vienna, 1781) as op.2 no.2	20, 44
304	300*c*	Sonata	e	Paris, early sum. 1778	XVIII/ii, 54	VIII:23/i, 98	(Paris, 1778) as op.1 no.4	42

K	K^6	Title	Key	Composition	MW	NMA	Remarks	
306	300*l*	Sonata	D	Paris, sum. 1778	XVIII/ii, 76	VIII:23/i, 118	(Paris, 1778) as op.1 no.6	42
378	317*d*	Sonata	B♭	Salzburg, 1779–80	XVIII/ii, 140	VIII:23/i, 154	(Vienna, 1781) as op.2 no.4	17, 20, 44
372	372	Sonata	B♭	Vienna, 24 March 1781	XVIII/ii, 98	VIII:23/ii, 154	Allegro only, inc.; completed by M. Stadler	
379	373*a*	Sonata	G	Vienna, April 1781	XVIII/ii, 160	VIII:23/ii, 3	(Vienna, 1781) as op.2 no.5	20, 44
359	374*a*	Variations	G	Vienna, June 1781	XVIII/ii, 290	VIII:23/ii, 136	on La bergère Célimène, Fr. song (anon.) (Vienna, 1786)	
360	374*b*	Variations	g	Vienna, June 1781	XVIII/ii, 300	VIII:23/ii, 144	on Hélas, j'ai perdu mon amant, Fr. song (anon.) (Vienna, 1786)	
376	374*d*	Sonata	F	Vienna, sum. 1781	XVIII/ii, 108	VIII:23/ii, 16	(Vienna, 1781) as op.2 no.1	20, 44
377	374*e*	Sonata	F	Vienna, sum. 1781	XVIII/ii, 124	VIII:23/ii, 32	(Vienna, 1781) as op.2 no.3	20, 44
380	374*f*	Sonata	E♭	Vienna, sum. 1781	XVIII/ii, 172	VIII:23/ii, 48	(Vienna, 1781) as op.2 no.6	20, 44
454	454	Sonata	B♭	Vienna, 21 April 1784 (V)	XVIII/ii, 210	VIII:23/ii, 64	(Vienna, 1784) as op.7 no.3	24, 44
481	481	Sonata	E♭	Vienna, 12 Dec 1785 (V)	XVIII/ii, 232	VIII:23/ii, 82	(Vienna, 1786)	
526	526	Sonata	A	Vienna, 24 Aug 1787 (V)	XVIII/ii, 252	VIII:23/ii, 100	(Vienna, 1787)	25, 44
547	547	Sonata	F	Vienna, 10 July 1788 (V)	XVIII/ii, 276	VIII:23/ii, 122	'für Anfänger'	

Frags.: KA46/374*g*, B♭, 33 bars, 1781–2; K403/385*c*, C, parts of 3 movts, 1784–5; K402/385*E*, A; K396/385*f*, C, 28 bars, *c*1781; KA48/385*e*, A, 34 bars, 1784–5; K404/485*d*, C, at least 24 bars, ?1786

Spurious: K55–60/C23.01–6, MW, XVIII, 114ff, see F. Neumann, *MJb 1965–6*, 152–60, Plath, D1968–70; K61, MW, XVIII, 172, by H.F. Raupach

KEYBOARD

Sonatas

K	K^6	Key	Composition	MW	NMA	Remarks	
			solo keyboard				
A199–202	33*d–g*	G, B♭, C, F	1766	—	—	lost; listed in Breitkopf catalogue	
279–83	189*d–h*	C, F, B♭, E♭, G	Munich, early 1775	XX, 1	IX:25		42
284	205*b*	D	Munich, Feb–March 1775	XX, 46	IX:25	(Vienna, 1784) as op.7 no.2; sketch in autograph	24, 42
309	284*b*	C	Mannheim, Oct–Nov 1777	XX, 64	IX:25	(Paris, 1782) as op.4 no.1	15, 41
311	284*c*	D	Mannheim, Nov 1777	XX, 92	IX:25	(Paris, 1782) as op.4 no.2	15, 41, 42
310	300*d*	a	Paris, sum. 1778	XX, 78	IX:25	(Paris, 1782) as op.4 no.3	41
330	300*h*	C	Munich or Vienna, 1781–3	XX, 106	IX:25	(Vienna, 1784) as op.6 no.1	19
331	300*i*	A	Munich or Vienna, 1781–3	XX, 118	IX:25	(Vienna, 1784) as op.6 no.2	19, 27
332	300*k*	F	Munich or Vienna, 1781–3	XX, 130	IX:25	(Vienna, 1784) as op.6 no.3	19
333	315*c*	B♭	Linz and Vienna, 1783–4	XX, 146	IX:25	(Vienna, 1784) as op.7 no.1	23, 24, 52
457	457	c	Vienna, 14 Oct 1784 (V)	XX, 160	IX:25	pubd with Fantasia K475 (Vienna, 1785) as op.11, see Wolf, O 1992	

K	K6	Key	Composition	MW	NMA	Remarks
533	533	F	Vienna, 3 Jan 1788 (V)	XXII, 44	IX:25	incl. rev. of Rondo K494; (Vienna, 1788)
545	545	C	Vienna, 26 June 1788 (V)	XX, 174	IX:25	'für Anfänger'
A135	547a	F	? Vienna, sum. 1788	—	IX:26, 157	doubtful; finale = transposed version of K545, iii
570	570	B♭	Vienna, Feb 1789 (V)	XX, 182	IX:25	first edn (1796) with vn acc., probably spurious
576	576	D	Vienna, July 1789 (V)	XX, 194	IX:25	53

Frags.: K400/372*a*, B♭, MW, XXIV, 26, NMA, IX:25/ii, 174; KA31/569*a*, B♭, ?1789, NMA, IX:25/ii, 181, ? related to K570; KA29, 30, 37/590*a–c*, F, 1789–90; K312/590*d*, g, ?1789–90, MW, XX, 13, NMA, IX:25/ii, 184; K—, C, *c*1773, NMA, IX:25/ii, 173

K	K6	Key	Composition	MW	NMA	Remarks
			keyboard duet			
381	123*a*	D	Salzburg, mid-1772	XIX, 32	IX:24/2, 20	(Vienna, 1783) as op.3 no.1
358	186*c*	B♭	Salzburg, late 1773 – early 1774	XIX, 18	IX:24/2, 36	(Vienna, 1783) as op.3 no.2
497	497	F	Vienna, 1 Aug 1786 (V)	XIX, 46	IX:24/2, 54	(Vienna, 1787) as op.12
521	521	C	Vienna, 29 May 1787 (V)	XIX, 80	IX:24/2, 106	(Vienna, 1787)
			for 2 keyboards			
448	375a	D	Vienna, Nov 1781	XIX, 126	IX:24/1, 2	

Frags.: KA42/375*b*, 1782–3; KA43/375*c*, B♭, 2 kbd, 1782–3; KA45/375*d*, G, 2 kbd, ?1785–6; KA44/426*a*, 2 kbd, ?1785–6; ? Sonata, G, kbd 4 hands [K357/497*a*, Allegro, 98 bars, and K357/500*a*, Andante, 158 bars], 1791, MW, XIX, 2, 10, NMA, IX:24/2, 142

Doubtful: K19*d*, C, NMA, IX:24/2, 2 (Paris, 1788), see Eisen, O 1998

Variations

K	K6	Theme	Key	Composition	MW	NMA	Remarks
				solo keyboard			
A206	21*a*	?orig.	C	? London, 1765	—	—	lost; listed in Breitkopf catalogue
24	24	Dutch song (Laat ons juichen) by C.E. Graaf	G	The Hague, Jan 1766	XXI, 1	IX:26, 3	(The Hague, 1766)
25	25	Willem van Nassau (Dutch national song)	D	Amsterdam, Feb 1766	XXI, 6	IX:26, 9	(The Hague, 1766)
180	173*c*	Mio caro Adone from Salieri: La fiera di Venezia, Vienna, 1772	G	Vienna, aut. 1773	XXI, 22	IX:26, 15	(Paris, 1778)
179	189*a*	Minuet [finale of Ob Conc. no.1, 1768] by J.C. Fischer	C	Salzburg, sum. 1774	XXI, 12	IX:26, 20	(Paris, 1778)
354	299*a*	Je suis Lindor (song in Beaumarchais: Le barbier de Séville, by A.L. Baudron)	E♭	Paris, early 1778	XXI, 58	IX:26, 34	(Paris, 1778)

K	K6	Theme	Key	Composition	MW	NMA	Remarks	
265	300*e*	Ah vous dirai-je, maman (Fr. song)	C	Vienna, 1781–2	XXI, 36	IX:26, 49	(Vienna, 1785)	
353	300*f*	La belle françoise (Adieu donc, dame françoise, Fr. song)	E♭	Vienna, 1781–2	XXI, 50	IX:26, 58	(Vienna, 1786)	
264	315*d*	Lison dormait from N. Dezède: Julie, Paris, 1772	C	Paris, late sum. 1778	XXI, 26	IX:26, 67	shortened (Paris, 1786), (Vienna, 1786)	
352	374*c*	Dieu d'amour (March), chorus from A.-E.-M. Grétry: Les mariages samnites, Paris, 1776	F	Vienna, June 1781	XXI, 44	IX:26, 82	(Vienna, 1786)	
398	416*e*	Salve tu, Domine, chorus from G. Paisiello: I filosofi immaginari, Vienna, 1781	F	Vienna, March 1783	XXI, 68	IX:26, 90	(Vienna, 1786)	
460	454*a*	Come un agnello from Sarti: Fra i due litiganti, Milan, 1782	A	Vienna, ? June 1784	XXI, 84	IX:26, 154	autograph has 2 variations; version with 8 variations (Vienna, 1784) probably by Sardi, see R. Armbruster, *MJb 1997*, 225–48	
455	455	Les hommes pieusement (Unser dummer Pöbel meint) from Gluck: La rencontre imprévue	G	Vienna, 25 Aug 1784 (V)	XXI, 74	IX:26, 98	(Vienna, 1785); earlier version ?1781–2	
500	500	probably orig.	B♭	Vienna, 12 Sept 1786 (V)	XXI, 94	IX:26, 112		
54	547*b*	probably orig.	F	Vienna, July 1788		IX:26, 157	1st edn (1785) has spurious 4th variation; re-used by Mozart, with vn, K547	
573	573	Minuet [from Vc Sonata op.4 no.6] by J.P. Duport	D	Potsdam, 29 April 1789 (V)	XXI, 100	IX:26, 120	(Berlin, 1791); see K. Hortschansky, *Mf*, xvi (1963), 265–7	
613	613	Ein Weib ist das herrlichste Ding, by B. Schack or F.X. Gerl	F	Vienna, March 1791	XXI, 108	IX:26, 132	theme from music to Schikaneder play Der dumme Gärtner aus dem Gebirge, 1789;(Vienna, 1791)	53

Frags.: KA38/383*d*, ?org, MW, XXII, 15, NMA, IX:26, 149; K236/588*b*, E♭, theme by Gluck, 1782–3, ? intended for variations

Doubtful: KA206/21*a*, ? London, 1764–5, lost

piano duet

K	K6	Theme	Key	Composition	MW	NMA	Remarks
501	501	probably orig.	G	Vienna, 4 Nov 1786 (V)	XIX, 108	IX:24/ii, 96	

Miscellaneous

K	K6	Title	Key	Composition	MW	NMA	Remarks	
				solo keyboard				
—	1*a*	Andante	C	Salzburg, early 1761	—	—		2
—	1*b*	Allegro	C	Salzburg, early 1761	—	—		2
—	1*c*	Allegro	F	Salzburg, 11 Dec 1761	—	—		
—	1*d*	Minuet	F	Salzburg, 16 Dec 1761	—	—		
1	1*e*	Minuet	G	Salzburg, Dec 1761 – Jan 1762	XII, 2	—		
—	1*f*	Minuet	C	Salzburg, Dec 1761 – Jan 1762	—	—		
2	2	Minuet	F	Salzburg, Jan 1762	XXII, 3	—		2
3	3	Allegro	B♭	Salzburg, 4 March 1762	XXII, 38	—		2
4	4	Minuet	F	Salzburg, 11 May 1762	XXII, 3	—		
5	5	Minuet	F	Salzburg, 5 July 1762	XXII, 4	—		2
9*a*	5*a*	Allegro	C	sum. 1763	—	—		
9*b*	5*b*	Andante	B♭	sum. 1763	—	—		
—	33*B*	[without title]	F	Zürich, Oct 1766	—	—		
41*e*	41*e*	Fugue		Salzburg, 1767	—	—	lost; in LC	
72*a*	72*a*	Allegro	G	? Verona, Jan 1770	—	—	inc.; only source is portrait by S. dalla Rosa	9
94	73*h*	Minuet	D	Salzburg, 1769	XXII, 5	—		
284*a*	284*a*	4 preludes			—	—	identical with K395/300*g*	
284*f*	284*f*	Rondo		Mannheim, Nov 1777	—	—	lost; mentioned in letter, 29 Nov 1777	
395	300*g*	Capriccio	C	Munich, Oct 1777	XXIV, no.24	—		
315*a*	315*g*	8 minuets		Salzburg, late 1773	—	—		
400	372*a*	Allegro	B♭	Vienna, 1781	XXIV, no.26	—	inc.; completed by M. Stadler	
401	375*e*	Fugue	g	Vienna, early 1782	XXII, 34	—	inc.; completed by M. Stadler; also duet version	
153	375*f*	Fugue	E♭	? Salzburg, 1783	XXIV, no.25	—	inc.; completed by S. Sechter	
394	383*a*	Prelude and fugue	C	Vienna, early 1782	XX, 20	—		
396	385*f*	Fantasia	c	Vienna, early 1782	XX, 214	IX:25	inc.; orig. with vn, see 'Chamber music'	
397	385*g*	Fantasia	d	Vienna, early 1782 or 1786–7	XX, 220	IX:25	last 10 bars (not in 1st edn) probably spurious; see Plath, in Plath and others, D 1971–2, 31	
399	385*i*	Suite	C	Vienna, early 1782	XXII, 28	—	Sarabande inc.	
154	385*k*	Fugue	g	Vienna, early 1782	XXIV	—	inc.	
453*a*	453*a*	Funeral march	c	Vienna, 1784	—	—		
475	475	Fantasia	c	Vienna, 20 May 1785 (V)	XX, 224	IX:25	pubd with Sonata K457 (Vienna, 1785) as op.11	
485	485	Rondo	D	Vienna, 10 Jan 1786 (D)	XXII, 8	IX:25	(Vienna, *c*1786)	
494	494	Rondo	F	Vienna, 10 June 1786 (D)	XXII, 14	IX:25	(London, 1788), (Speyer, 1788); rev. version in Sonata K533	
511	511	Rondo	a	Vienna, 11 March 1787 (V, D)	XXII, 20	IX:25	(Vienna, 1787)	
540	540	Adagio	b	Vienna, 19 March 1788 (V)	XXII, 56	—	? (Vienna, 1788)	
574	574	Gigue	G	Leipzig, 16 May 1789 (D)	XXII, 60	—		

K	K[6]	Title	Key	Composition	MW	NMA	Remarks	
355	576*b*	Minuet	D	Vienna, ?1786–7	XXII, 6	—	trio by M. Stadler; see King, B 1955, 3/1970, 222–3; Badura-Skoda, *NZM*, Jg.127 (1966), 468–72	
236	588*b*	Andantino	E♭		XXII, 55	—	see 'Arrangements'	
312	590*d*	Allegro	g	Vienna, 1789–90	XXII, 39	—	inc.; ? for a sonata; see W. Plath, in Plath and others, D 1971–2, 30–31; Tyson, D 1987, 20	
—	—	[without title]	E♭	? Salzburg, Jan 1769	—	—		

Frags.: K73*w*, Fugue, D, early 1773; KA41/375*g*, Fugue, G, 1777; K375*b*, Fugue, F; KA433 and 40/383*b*, Fugue, F, ?1788–9; KA39/383*d*, Fugue, c; KA32/383*C*, Fantasia, f; KA34/385*b*, Adagio, d, 1786–7; KA34/576*a*, Minuet, D, 1786–7; K—, untitled, B♭, ? Salzburg, 1769, see K626*b*/25

2 keyboards

K	K[6]	Title	Key	Composition	MW	NMA	Remarks	
426	426	Fugue	c	Vienna, 29 Dec 1783 (D)	XIX, 118	IX:24/1, 39	(Vienna, 1788); arr., with new introduction, for str, K546	45
—	—	Larghetto and Allegro	E♭	? Vienna, 1782–3	—	IX:24/1, suppl.	inc.; completed by M. Stadler; see G. Croll, *MJb 1962–3*, 108–10; *MJb 1964*, 28–37	

Frags.: KA42/375*b*, Grave–Presto, B♭, 52 bars, MW, XXIV, 60, NMA, IX:24/1, 46; KA43/375*c*, B♭, 15 bars, NMA, IX:24/1, 49; KA45/375*d*, Fugue, G, 23 bars, NMA, IX:24/1, 50; KA44/426*a*, Allegro, c, 22 bars, NMA, IX:24/1, 51

for mechanical organ or armonica

K	K[6]	Title	Instrument	Key	Composition	MW	NMA	Remarks	
594	594	Adagio and Allegro	mechanical org	f	Vienna and elsewhere, Oct–Dec 1790	XXIV, no.27a	IX:27		30
608	608	[Fantasia]	mechanical org	f	Vienna, 3 March 1791 (C)	X/100	IX:27		30
616	616	Andante	mechanical org	F	Vienna, 4 May 1791	X/109	IX:27	arr. pf (Venice, 1791)	30
356	617*a*	Adagio	armonica	C	Vienna, 1791	X/84	IX:27		

Frags. for mechanical org: KA35/593*a*, Adagio, d, 1790–91; K615*a*, Andante, F, 1791

MISCELLANEOUS

K	K6	Title	Key	Composition	MW	NMA	Remarks
A109*b*, 15*a–ss*	15 *a–ss*	London Sketchbook		London, 1765	—	—	short pieces on 2 staves for kbd or sketches for orch
—	32a	Capricci		?1764–6	—	—	lost; see C. Mozart's letter to André, 2 March 1799; ? in LC
41*f*	41*f*	Fugue a 4		Salzburg, 1767	—	—	lost; in LC
393	385*b*	Solfeggios for voice		Vienna, ?Aug 1782	XXIV, no.49	—	
—	453*b*	Exercise book for Barbara Ployer			—	—	facs. in R. Lach, *W.A. Mozart als Theoretiker* (Vienna, 1918)
485*a*	506*a*	Attwood Studies		Vienna, 1785–6	—	X:30, 1	
A294*d*	516*f*	Musikalisches Würfelspiel	C	Vienna, 1787	—	—	
A78	620*b*	[contrapuntal study]	b	Vienna, ? Sept 1791	—	—	chorale setting; ? sketch for Die Zauberflöte K620

Frags.: K—/385*n*, Fugue a 4, A, Vienna, ?1782; K443/404*b*, Fugue a 3, G, Vienna, ?1782, completed by M. Stadler

Doubtful: K154/A61–2, fugues, before 1772; K—/A65, Adagio, F, ed. N. Zaslaw in *Haydn, Mozart and Beethoven: Essays in Honour of Alan Tyson*, ed. S. Brandenburg (Oxford, 1998), 101–14

ARRANGEMENTS ETC.

K	K6	Orig. composer, work	Orig. scoring	Key	Mozart's scoring	Date of arr.	MW	NMA	Remarks	
37	37	i Raupach, op.1 no.5 ii ? iii L. Honauer, op.2 no.3	kbd	F	kbd, 2 ob, 2 hn, str	Salzburg, April 1767	XVI/i, 1	X:28/ii, 3		6, 36
39	39	i Raupach, op.1 no.1 ii J. Schobert, op.17 no.2 iii Raupach, op.1 no.1	kbd	B♭	kbd, 2 ob, 2 hn, str	Salzburg, June 1767	XVI/i, 35	X:28/ii, 45		6, 36
40	40	i Honauer, op.2 no.1 ii J.G. Eckard, op.1 no.4 iii C.P.E. Bach, H81 W117	kbd	D	kbd, 2 ob, 2 hn, str	Salzburg, July 1767	XVI/i, 67	X:28/ii, 84	cadenza K624/626*a*II, *C*	6, 36
41	41	i Honauer, op.1 no.1 ii Raupach, op.1, no.1 iii Honauer, op.1 no.1	kbd	G	kbd, 2 ob, 2 hn, str	Salzburg, July 1767	XVI/i, 99	X:28/ii, 125		6, 36
104	61*e*	M. Haydn, minuets	orch	C, F, C, A, G, G	orch	Salzburg, *c*1771	—	IV:13/1/i, 28		8
—	61*g*II	M. Haydn, minuet	kbd	C	orch	Salzburg, *c*1771	—	IV:13/1/i, 92		
122	73*t*	M. Haydn, minuet	orch	E♭	orch	? Bologna, Aug 1770	XXIV, no.13a	IV:13/1/i, 10		
44	73*u*	J. Stadlmayr, Musica super cantum gregorianum	5vv		SATB	Salzburg, *c*1768–9	—	—	see E. Hintermaier, *MJb 1991*, 509–17	
107, 1	107, 1	J.C. Bach, op.5 no.2	kbd	D	kbd, 2 vn, b	1772	—	X:28/ii, 165	cadenzas K624/626*a*II, *A–B*	36
107, 2	107, 2	J.C. Bach, op.5 no.3	kbd	G	kbd, 2 vn, b	1772	—	X:28/ii, 187		
107, 3	107, 3	J.C. Bach, op.5 no.4	kbd	E♭	kbd, 2 vn, b	1772	—	X:28/ii, 203		

K	K⁶	*Orig. composer, work*	*Orig. scoring*	*Key*	*Mozart's scoring*	*Date of arr.*	*MW*	*NMA*	*Remarks*	
284*e*	284*e*	J.B. Wendling, conc.	fl, str		?addl wind	Mannheim, Nov 1777	—	—	lost; see letter, 21 Nov 1777	
404*a*	404*a*	6 preludes and fugues	kbd		vn, va, vc	Vienna, 1782	—	—	doubtful; see Kirkendale, N 1964 and Kirkendale, *Mf*, xviii (1965), 195–9; Holschneider, *Mf*, xvii (1964), 51–6	
		1 p ?orig., f J.S. Bach BWV853		d						
		2 p ?orig., f BWV883		g						
		3 p ?orig., f BWV882		F						
		4 p BWV527/ii, f BWV1080 no.8		F						
		5 p, f BWV526/ii, iii		E♭						
		6 p ?orig., f W.F. Bach Fugue no.8		f						
405	405	J.S. Bach, 5 fugues BWV871, 876, 878, 877, 874	kbd	c, E♭, E, d, D	2 vn, va, vc	Vienna, 1782	—	—	see W. Kirkendale, *MJb 1962–3*, 140–55	
—	—	J.S. Bach, BWV891	kbd	c	2 vn, va, vc	? Vienna, 1782	—	—	see G. Croll, *ÖMz*, xxi (1966), 508–14	
—	—	6 preludes and fugues	kbd		2 vn, va, vc	? Vienna, 1782	—	—	very doubtful; see Kirkendale, N 1964	
		1 p ?orig., f J.S. Bach BWV548		e						
		2 p ?orig., f BWV877		d						
		3 p ?orig., f BWV876		E♭						
		4 p ?orig., f BWV891		b						
		5 p ?orig., f BWV874		D						
		6 p ?orig., f BWV878		E						
—	—	3 preludes and fugues	kbd		2 vn, 2 va, vc	? Vienna, 1782	—	—	very doubtful; see Kirkendale, N 1964	
		1 p ?orig., f J.S. Bach BWV849		d						
		2 p ?orig., f BWV867		a						
		3 p ?orig., f BWV546		c						
470*a*	470*a*	G.B. Viotti, Vn Conc. no.16			addl tpt, timp	Vienna, *c*1789–90	—	—	see M.H. Schmid, *Mozart-Studien*, v (1995), 149–71	
—	506*a*, HS4	J. Haydn, duet Cara, sarò fedele, from Armida				Vienna, *c*1786–91	—	—	formerly considered part of the Attwood exercises; facs. in Landon, G 1989	
A109*g* no.19	537*d*	C.P.E. Bach, Ich folge dir, from Auferstehung und Himmelfahrt Jesu	T, tpt, str		addl fl, ob, tpt	Vienna, Feb 1788	—	—		
566	566	G.F. Handel, Acis and Galatea	S, T, T, T, B, rec, 2 ob, bn, 2 vn, va, bc		addl 2 fl, 2 cl, bn, 2 hn	Vienna, Nov 1788	—	X:28/1/i		30
572	572	Handel, Messiah	S, A, T, B, SATB, 2 ob, 2 tpt, timp, str		addl 2 fl, 2 cl, 2 bn, 2 hn, 3 trbn, rev. tpt parts	Vienna, March 1789	—	X:28/1/ii		30

K	K6	Orig. composer, work	Orig. scoring	Key	Mozart's scoring	Date of arr.	MW	NMA	Remarks	
591	591	Handel, Alexander's Feast	S, T, B, SATB, 2 rec, 2 ob, 3 bn, 2 hn, 2 tpt, timp, str		addl 2 fl, 2 cl, rev. tpt parts	Vienna, July 1790	—	X:28/1/iii		30
592	592	Handel, Ode for St Cecilia's Day	S, T, SATB, fl, 2 ob, 2 tpt, timp, lute, str		addl fl, 2 cl, 2 bn, 2 hn, rev. tpt parts	Vienna, July 1790	—	X:28/1/iv		30
625	592*a*	Nun liebes Weibchen	—		—	—	—	—	see 'Duets and Ensembles for Solo Voices and Orchestra'	
624	626*a*II, *D–0*	Cadenzas	kbd			various	—	—	*D*(A61a), *F–G*, *H* for Schroeter op.3 nos.1, 4, 6; *K* for I. von Beecke, Conc. in D; *N*, *0* for unknown conc; *L* lost; *E*, *I* unauthentic	
	626*b*, 28	Gluck, gavotte from Paride ed Elena	orch		2 fl, 5 tpt, timp		—	—	? Mozart's contribution to Divertimento K187/C17.12	
18	A51	C.F. Abel, Sym. op.7 no.6	orch	E♭	addl cls	London, 1764–5	—	—	edn (Basle, 1976)	
—	—	L. Mozart, Litaniae de venerabili altaris sacramento	S, A, T, B, SATB, 2 hn, str	D	various changes		—	X:28/3–5/i		
—	—	L. Mozart, litany	SATB, orch	E♭	trbn/va solo arr. for ob	Salzburg, ?*c*1774	—	X:28/3–5/ii	see Eisen, D 1991, 287–9	
—	—	L. Mozart, litany	SATB, orch	D	various changes, esp. to hn part	Vienna, 1781–2	—	—		

K293*e*, 19 cadenzas for arias by J.C. Bach and others (unidentified): see Plath, D 1960–61, 106, and in Plath and others, D 1971–2, 20

Frag.: K—, Handel, Fugue, kbd, F, HWV427, Vienna, 1782–3

Bibliography

A: Catalogues, bibliographies, letters, documents, iconography. B: Compendia, collective works, congress reports, periodicals. C: Exhibition catalogues. D: Sources, authenticity, chronology, editions. E: Sketches, fragments, compositional process. F: Biographies, studies of life and works. G: Life: particular aspects and episodes. H: Works: style, influences, particular aspects. I: Sacred works. J: Operas. K: Arias, songs and other vocal music. L: Symphonies, serenades, etc. M: Concertos. N: Chamber music. O: Keyboard music. P: Performing practice. Q: Reception.

A: Catalogues, bibliographies, letters, documents, iconography

L. von Köchel: *Chronologisch-thematisches Verzeichnis sämtlicher Tonwerke Wolfgang Amadé Mozarts* (Leipzig, 1862; rev. 2/1905 by P. Graf von Waldersee; rev. 3/1937 by A. Einstein, repr. 4/1958, 5/1963, with suppl. 3/1947; rev. 6/1964 by F. Giegling, A. Weinmann and G. Sievers, repr. 7/1965)

C. von Wurzbach: *Mozart-Buch* (Vienna, 1869) [repr. of articles in C. von Wurzbach: *Biographisches Lexikon des Kaiserthums Oestereich* (Vienna, 1856–91)]

H. de Curzon: *Essai de bibliographie mozartienne: revue critique des ouvrages relatifs à W.A. Mozart et ses oeuvres* (Paris, 1906)

L. Schiedermair, ed.: *Die Briefe W.A. Mozarts und seiner Familie* (Munich, 1914)

O. Keller: *Wolfgang Amadeus Mozart: Bibliographie und Ikonographie* (Berlin, 1927)

R. Tenschert: *Wolfgang Amadeus Mozart 1756–1791: sein Leben in Bildern* (Leipzig, 1935)

E. Anderson, ed.: *The Letters of Mozart and his Family* (London, 1938; rev. 2/1966 by A.H. King and M. Carolan; rev. 3/1985 by S. Sadie and F. Smart)

E. Müller von Asow, ed.: W.A. Mozart: *Verzeichnis aller meiner Werke* (Vienna, 1943, 2/1956 with L. Mozart: *Verzeichnis der Jugendwerke W.A. Mozarts*)

R. Bory: *La vie et l'oeuvre de Wolfgang-Amadeus Mozart par l'image* (Geneva, 1948) [also Eng. edn]

O.E. Deutsch: 'Mozart's Portraits', in Landon and Mitchell, B 1956, 1–9

O.E. Deutsch: *Mozart: die Dokumente seines Lebens, gesammelt und erläutert* (Kassel, 1961; suppl., 1978, ed. J.H. Eibl; Eng. trans., 1965/*R*)

M. Zenger and O.E. Deutsch: *Mozart und seine Welt in zeitgenössischen Bildern/Mozart and his World in Contemporary Pictures* (Kassel, 1961)

W.A. Bauer, O.E. Deutsch and J.H. Eibl, eds.: *Mozart: Briefe und Aufzeichnungen* (Kassel, 1962–75) [complete edn; for later discoveries see C.B. Oldman, *MR*, xvii (1956), 68–9; W. Rehm, *Festschrift Rudolf Elvers*, ed. F. Herttrich and H. Scheider (Tutzing, 1985), 418–19; A. Goldmann, *Acta mozartiana*, xxxiv (1987), 62–3; J. Mančal, ibid., 77–82; J. Mančal, *ÖMz*, xlii (1987), 290–91; A. Briellmann, *MJb 1987–8*, 233–48; T. Leibnitz and A. Ziffer, *Katalog der Sammlung Anton Dermota* (Tutzing, 1988), 106–8; R. Angermüller, *Festschrift Wolfgang Rehm*, ed. D. Berke and H. Heckmann (Kassel, 1989), 155–66; Volek and Bittner, A 1991, 36–8, 138–44; U. Walter, *Mozart in Kursachsen*, ed. B. Richter and U. Oehme (Leipzig, 1991), 133–44]

R. Angermüller and O. Schneider: 'Mozart-Bibliographie (bis 1970)', *MJb 1975* [whole issue; with 5-year suppls.]

A. Hutchings: *Mozart: the Man, the Musician* (London, 1976)

H.-G. Klein: *Wolfgang Amadeus Mozart, Autographe und Abschriften: Katalog* (Laaber, 1982)

N. Zaslaw: 'Leopold Mozart's List of his Son's Works', *Music in the Classic Period: Essays in Honor of Barry S. Brook*, ed. A.W. Atlas (New York, 1985), 323–74

G. Haberkamp: *Die Erstdrucke der Werke von Wolfgang Amadeus Mozart: Bibliographie* (Tutzing, 1986)

A. Tyson and A. Rosenthal, eds.: *Mozart's Thematic Catalogue* (London, 1990; Ger. trans., Neue Ausgabe sämtlicher Werke, X:33/i, Kassel, 1991) [incl. facs.]

C. Eisen: *New Mozart Documents* (London and Stanford, SA, 1991); in Ger. with addns, as *Mozart: die Dokumente seines Lebens: Addenda, Neue Folge*, Neue Ausgabe sämtlicher Werke, X:31/ia (Kassel, 1997) [further suppl. to Deutsch, A 1961]

N. Zaslaw and F.M. Fein, eds.: *The Mozart Repertory: a Guide for Musicians, Programmers and Researchers* (Ithaca, NY, 1991)

A. Tyson, ed.: *Wasserzeichen-Katalog*, Neue Ausgabe sämtlicher Werke, X:33/ii (Kassel, 1992)

T. Volek and I. Bittner: *Mozartovské stopy v českych a moravskych archivech* (Prague, 1991); Eng. trans. as *The Mozartiana of Czech and Moravian Archives* (Prague, 1991)

R. Angermüller: 'Leopold Mozarts Verlassenschaft', *MISM*, xli/3–4 (1993), 1–32

I. Sorensen: 'Ein Mozart-Porträt in Dänemark', *MJb 1994*, 79–88

J. Mančal: *Mozartschätze in Augsburg* (Augsburg, 1995)

C. Eisen: 'The Mozarts' Salzburg Music Library', in Eisen, B 1997, 85–138

B: Compendia, collective works, congress reports, periodicals

Mitteilungen für die Mozart-Gemeinde in Berlin (1895–1925)

Mozarteums-Mitteilungen (1918–21)

Mozart-Jb 1923–9

Bulletin de la Société d'études Mozartiennes, i (1930–32)

Musikwissenschaftliche Tagung der Internationalen Stiftung Mozarteum: Salzburg 1931

Wiener Figaro: Mitteilungen der Mozartgemeinde Wien (1931–54); contd as *Mozartgemeinde Wien* (1954–85)

Neues Mozart-Jb 1941–3

H.F. Deininger, ed.: *Augsburger Mozartbuch* (Augsburg, 1942–3)

E.F. Schmid: *Ein schwäbisches Mozartbuch* (Lorch and Stuttgart, 1948)

MJb 1950– (1952–)

Acta mozartiana (1954–)

Prefaces and Critical Commentaries to all vols of *Wolfgang Amadeus Mozart: Neue Ausgabe sämtlicher Werke* (Kassel, 1955–91)

A.H. King: *Mozart in Retrospect: Studies in Criticism and Bibliography* (London, 1955, 3/1970/*R*)

Leben und Werk W.A. Mozarts: Prague 1956 (Prague, ?1958)

Kongress Musikwissenschaftlicher: Vienna 1956 (Graz, 1958)

Les influences étrangères dans l'oeuvre de W.A. Mozart: Paris 1956 (Paris, 1958)

A. Einstein: *Essays on Music* (New York, 1956, 2/1958) [incl. 8 on Mozart]

H.C.R. Landon and D. Mitchell, eds.: *The Mozart Companion* (London and New York, 1956/*R*)

P. Schaller and H. Kühner, eds.: *Mozart: Aspekte* (Olten, 1956) [symposium]
H.F. Deininger, ed.: *Neues Augsburger Mozartbuch* (Augsburg, 1962)
F. Blume *Syntagma musicologicum: gesammelte Reden and Schriften*, i, ed. M. Ruhnke (Kassel, 1963) ['Haydn und Mozart', 571–82; 'Wolfgang Amadeus Mozart', 583–669; 'Wolfgang Amadeus Mozart: Geltung und Wirkung', 670–86; 'Mozarts Konzerte und ihre Überlieferung', 686–714; 'Requiem und kein Ende', 714–34]
P.H. Lang, ed.: *The Creative World of Mozart* (New York, 1963/*R*)
Mozartgemeinde Wien, 1913–1963: Forscher und Interpreten (Vienna, 1964)
Idomeneo Conference: Salzburg 1973 [*MJb 1973–4*]
E. Wellesz and F. Sternfeld, eds.: *The Age of Enlightenment, 1745–1790*, NOHM, vii (1973)
Mozart und seine Unwelt: Salzburg 1976 [*MJb 1978–9*]
G. Weiss, ed.: *Festschrift Erich Valentin* (Regensburg, 1976) [incl. 12 essays on Mozart]
Mozart und Italien: Rome 1974 [*AnMc*, no.18 (1978)]
C. Roleff, ed.: *Collectanea mozartiana* (Tutzing, 1988)
P. Csobádi, ed.: *Wolfgang Amadeus: summa summarum. Das Phänomen Mozart: Leben, Werk* (Vienna, 1990)
H.C.R. Landon, ed.: *The Mozart Compendium: a Guide to Mozart's Life and Music* (London, 1990)
N. Zaslaw and W. Cowdery, eds.: *The Compleat Mozart* (New York, 1990)
Mozart e i musicisti italiani del suo tempo: Rome 1991 (Rome. 1994)
Mozart: origines et transformations d'un mythe: Clermont-Ferrand 1991 (Berne, 1994)
Mozart und Mannheim: Mannheim 1991 (Frankfurt, 1994)
Musikwissenschaftlicher Kongress zum Mozartjahr: Baden, nr Vienna, 1991 (Vienna, 1994)
On Mozart: Washington DC 1991 (Wahsington DC, 1994)
Wolfgang Amadeus Mozart: Leipzig 1991 (Leipzig, 1993)
P. Elder and G. Walterskirchen, eds.: *Das Benediktinerstift St. Peter in Salzburg zur Zeit Mozarts* (Salzburg, 1991)
C. Eisen, ed.: *Mozart Studies* (Oxford, 1991)
W.A. Mozart in Wien und Prague: die grossen Opern: Vienna 1992 (Vienna, 1992)
S. Sadie, ed.: *Wolfgang Amadè Mozart: Essays on his Life and Music* (Oxford, 1996)
C. Eisen, ed.: *Mozart Studies 2* (Oxford, 1997)

C: Exhibition catalogues

Mozart en France, Bibliothèque Nationale (Paris, 1956)
F. Hadamowsky and L. Nowak, eds.: *Mozart: Werk und Zeit*, Österreichische Nationalbibliothek, 30 May – 30 Sept 1956 (Vienna, 1956)
R. Münster, ed.: *La finta giardiniera: Mozarts Münchner Aufenthalt, 1774/75*, Bayerische Staatsbibliothek, 13 Jan – 28 Feb 1975 (Munich, 1975)
R. Münster, ed.: *Wolfgang Amadeus Mozart: Idomeneo, 1781–1981*, Bayerische Staatsbibliothek, 27 May – 31 July 1981 (Munich, 1981)
R. Angermüller, ed.: *'Auf Ehre und Credit': die Finanzen des W.A. Mozart* (Munich, 1983)
T. Volek and J. Pesčková, eds.: *Mozartuv Don Giovanni: výstava k 200. výročí svečtové premiéry v Praze 1787–1987* [exhibition on the 200th anniversary of the world première in Prague] (Prague, 1987)

R. Klein and F. Zorrer, eds.: *Bruder Wolfgang Amadeus Mozart: 7. Sonderausstellung 1990–91 des Österreichischen Freimaurermuseums Schloss Rosenau bei Zwettl* (Vienna, 1990)
R. Angermüller, ed.: *Mozart, Bilder und Klänge: Salzburger Landesausstellung Schloss Klessheim*, 6 March – 3 Nov 1991 (Salzburg, 1991)
C.A. Banks and J.R. Turner: *Mozart: Prodigy of Nature* (London and New York, 1991)
H.-G. Klein and H. Hell: *Wolfgang Amadeus Mozart: 'Componieren, meine einzige Freude und Passion': Autographe und frühe Drucke aus dem Besitz der Berliner Staatsbibliothek*, 5 Dec 1991 – 8 Feb 1992 (Wiesbaden, 1991)
G. Brosche, ed.: *Requiem, Wolfgang Amadeus Mozart 1791–1991: Ausstellung der Musiksammlung der Österreichischen Nationalbibliothek*, 17 May – 5 Dec 1991 (Graz, 1991)
U. Konrad and M. Staehelin, eds.: *Allzeit ein Buch: die Bibliothek Wolfgang Amadeus Mozarts*, Biblioteca Augusta, 5 Dec 1991 – 15 March 1992 (Weinheim, 1991)
A. Rohrmoser and J. Neuhardt, eds.: *Katalog zur Ausstellung: Salzburg zur Zeit der Mozart* (Salzburg, 1991)
A. Rosenthal and P. Ward Jones: *Mozart: a Bicentennial Loan Exhibition: Autograph Music Manuscripts, Letters, Portraits, First Editions and Other Documents of the Composer and his Circle*, Bodleian Library (Oxford, 1991)
N. Salinger and H.C.R. Landon, eds.: *Mozart à Paris*, Musée Carnavalet, 13 Nov 1991 – 16 Feb 1992 (Paris, 1991)
M. Schneider, ed.: *Mozart in Tirol: Ausstellungskatalog des Tiroler Landesmuseums Ferdinandeum Innsbruck*, 29 May – 29 Sept 1991 (Innsbruck, 1991)
G. Stradner, ed.: *Die Klangwelt Mozarts: Ausstellungskatalog des Kunsthistorischen Museums, Sammlung alter Musikinstrumente*, 28 April – 27 Oct 1991 (Vienna, 1991)
B. Tellini Santoni, ed.: *Il teatro di Mozart a Roma*, Bibliotheca Vallicelliana (Rome, 1991)
K. von Welck and L. Homering: *176 Tage W.A. Mozart in Mannheim*, Reiss-Museum, 19 Sept 1991 – 26 Jan 1992 (Mannheim, 1991)
R. Fuhrmann, ed.: *Mozart und die Juden*, Haus der Bremischen Bürgerschaft, 12 Oct – 11 Nov 1994 (Bremen, 1994)

D: Sources, authenticity, chronology, editions

L. Schiedermair: *W.A. Mozarts Handschrift in zeitlich geordneten Nachbildungen* (Bückeburg, 1919)
C.B. Oldman: 'Mozart and Modern Research', *PRMA*, lviii (1931–2), 43–66
O.E. Deutsch: 'Mozarts Nachlass: aus den Briefen Constanzes an den Verlag André', *MJb 1953*, 32–7
O.E. Deutsch: 'Mozarts Verleger', *MJb 1955*, 49–55
A.H. King: *Mozart in the British Museum* (London, 1956/*R*)
L. Nowak: 'Die Wiener Mozart-Autographen', *ÖMz*, xi (1956), 180–87
A. Weinmann: *Wiener Musikverleger und Musikalienhändler von Mozarts Zeit bis gegen 1860* (Vienna, 1956)
M. and C. Raeburn: 'Mozart's Manuscripts in Florence', *ML*, xl (1959), 334–40
L. Finscher: 'Maximilian Stadler und Mozarts Nachlass', *MJb 1960–61*, 168–72
W. Plath: 'Beiträge zur Mozart-Autographie I: die Handschrift Leopold Mozarts', *MJb 1960–61*, 82–118

A. Holschneider: 'Neue Mozartiana in Italien', *Mf*, xv (1962), 227–36

W. Senn: 'Die Mozart-Überlieferung im Stift Heilig Kreuz zu Augsburg', in Deininger, B 1962, 333–68

K.-H. Köhler: 'Die Erwerbung der Mozart-Autographe der Berliner Staatsbibliothek: ein Beitrag zur Geschichte des Nachlasses', *MJb 1962–3*, 55–68

W. Plath: 'Miscellanea Mozartiana I', *Festschrift Otto Erich Deutsch*, ed. W. Gerstenberg, J. LaRue and W. Rehm (Kassel, 1963), 135–40

W. Rehm: 'Miscellanea Mozartiana II', ibid., 141–54

H. Engel: 'Probleme der Mozartforschung', *MJb 1964*, 38–54

W. Plath: 'Der Ballo des "Ascanio" und die Klavierstücke KV Anh., 207', *MJb 1964*, 111–29

W. Plath: 'Der gegenwärtige Stand der Mozart-Forschung', *IMSCR IX: Salzburg 1964* (Kassel, 1964–6), i, 47–56; ii, 88–97

G. Rech: 'Ergebnisse der heutigen Erforschung seines Lebenswerkes', *Universitas* [Stuttgart], xix (1964), 925; Eng. trans. in *Universitas*, vii (1964–5), 355–60

H. Federhofer: 'Mozartiana im Musikaliennachlass von Ferdinand Bischoff', *MJb 1965–6*, 15–38

D. Kolbin: 'Autographe Mozarts und seiner Familie in der UdSSR', *MJb 1968–70*, 281–303

W. Plath: 'Mozartiana in Fulda und Frankfurt', *MJb 1968–70*, 333–86

M.H. Schmid, ed.: *Die Musiksammlung der Erzabtei St. Peter in Salzburg: Katalog I: Leopold und Wolfgang Mozart, Joseph und Michael Haydn* (Salzburg, 1970)

K. Pfannhauser: 'Epilegomena Mozartiana', *MJb 1971–2*, 268–312

W. Plath: 'Leopold Mozarts Notenbuch für Wolfgang (1762): eine Fälschung?', *MJb 1971–2*, 337–41

W. Plath and others: 'Echtheitsfragen', *MJb 1971–2*, 7–67 [incl. discussions]

N. Zaslaw: 'A Rediscovered Mozart Autograph at Cornell University', *MJb 1971–2*, 419–31

D.N. Leeson and D. Whitwell: 'Mozart's Thematic Catalogue', *MT*, cxiv (1973), 781–3

A.H. King: 'Some Aspects of Recent Mozart Research', *PRMA*, c (1973–4), 1–18

W. Senn: 'Beiträge zur Mozartforschung', *AcM*, xlviii (1976), 205–27

W. Plath: 'Beiträge zur Mozart-Autographie II: Schriftchronologie 1770–1780', *MJb 1976–7*, 131–73

A. Tyson: *Mozart: Studies of the Autograph Scores* (Cambridge, MA, 1987)

A. Tyson: 'A Feature of the Structure of Mozart's Autograph Scores', *Festschrift Wolfgang Rehm*, ed.: D. Berke and H. Heckmann (Kassel, 1989), 95–105

A. Tyson: 'Some Features of the Autograph Score of *Don Giovanni*', *Israel Studies in Musicology* (1990), 7–26

W. Plath: *Mozart-Schriften: ausgewählte Aufsätze*, ed. M. Danckwardt (Kassel, 1991)

C. Eisen: 'The Old and New Mozart Editions', *EMc*, xix (1991), 513–32

C. Eisen: 'The Mozarts' Salzburg Copyists: Aspects of Attribution, Chronology, Text, Style and Performance Practice', in Eisen, B 1991, 253–307

E. Warburton, ed.: *The Librettos of Mozart's Operas* (New York, 1992)

A. Tyson: 'Problems in Three Mozart Autographs in the Zweig Collection in the British Library', *Sundry Sorts of Music Books: Essays on the British Library Collections Presented to O.W. Neighbour*, ed. C. Banks, A. Searle and M. Turner (London, 1993), 248–55

E: Sketches, fragments, compositional process

M. Blaschitz: *Die Salzburger Mozart-Fragmente* (diss., U. of Bonn, 1924, part pubd in *Jb der Dissertationen der philosophischen Fakultät*, Bonn, 1924–5)

R. Engländer: 'Die Mozart-Skizzen der Universitätsbibliothek Uppsala: eine entstehungsgeschichtliche Studie', *STMf*, xxxvii (1955), 96–118; suppl. in *Mf*, ix (1956), 307–8

E. Hertzmann: 'Mozart's Creative Process', *MQ*, xliii (1957), 187–200; repr. in Lang, B 1963, 17–30

W. Plath: 'Das Skizzenblatt KV467a', *MJb 1959*, 114–26

W. Senn: 'Mozarts Skizze der Ballettmusik zu Le gelosie del serraglio', *AcM*, xxxiii (1961), 169–92

E. Hess: 'Ein neu entdecktes Skizzenblatt Mozarts', *MJb 1964*, 185–92

W. Plath: 'Bemerkungen zu einem missdeuteten Skizzenblatt Mozarts', *Festschrift Walter Gerstenberg*, ed. G. von Dadelsen and A. Holschneider (Wolfenbüttel, 1964), 143–50

W. Plath: 'Überliefert die dubiose Klavier-Romanze in As KV-Anh.205, das verschollene Quintett-Fragment KV-Anh.54 (452a)?', *MJb 1965–6*, 71–86

G. Croll: 'Zu den Verzeichnissen von Mozarts nachgelassenen Fragmenten und Entwurfen', *ÖMz*, xxi (1966), 250–54

C. Wolff: 'The Challenge of Blank Paper: Mozart the Composer', *On Mozart: Washington DC 1991* (Washington DC, 1994), 113–29

C. Wolff: 'Musikalische Gedanken und thematische Substanz: analytische Aspekte der Mozart-Fragmente', *MJb 1991*, 922–9

U. Konrad: *Mozarts Schaffensweise: Studien zu den Werkautographen, Skizzen und Entwürfen* (Göttingen, 1992) [incl. thematic catalogue]

U. Konrad: 'Neuentdecktes und wiedergefundenes Werkstattmaterial Wolfgang Amadeus Mozarts: Erster Nachtrag zum Katalog der Skizzen und Entwürfe', *MJb 1995*, 1–28 [suppl. to Konrad E 1992]

F: Biographies, studies of life and works

MGG1 (F. Blume; iconography by W. Rehm, bibliography and work-list by F. Lippmann, list of edns by R. Schaal)

F. Schlichtegroll: 'Johannes Chrysostomus Wolfgang Gottlieb Mozart', *Nekrolog auf das Jahr 1791* (Gotha, 1793), ed. L. Landshoff (Munich, 1924); as *Mozarts Leben* (Graz, 1794/*R*)

F.X. Niemetschek: *Leben des k.k. Kapellmeisters Wolfgang Gottlieb Mozart nach Originalquellen beschrieben* (Prague, 1798/*R*, enlarged 2/1808/*R*; Eng. trans., 1956/*R*); ed. J. Perfahl (Munich, 1984, 4/1991)

[I.F. Arnold] *Mozarts Geist: seine kurze Biografie und ästhetische Darstellung seiner Werke* (Erfurt, 1803)

I.T.F.C. Arnold: 'W.A. Mozart und Joseph Haydn: Nachträge zu ihren Biografien und ästhetischen Darstellung ihrer Werke: Versuch einer Parallele', *Gallerie der berühmtesten Tonkünstler des achtzehnten und neunzehnten Jahrhunderts* (Erfurt, 1810/*R*, 2/1816), i, 2–168

L.-A.-C. Bombet [Stendhal]: *Lettres ... sur le célèbre compositeur Haydn, suivies d'une vie de Mozart et considérations sur Métastase* (Paris, 1814, rev. 2/1817 as *Vies de Haydn, de Mozart et de Métastase*, 3/1854; Eng. trans., 1817, 2/1818, as *The Life of Haydn*; new Eng. trans., 1972, as *Lives of Haydn, Mozart and Metastasio*)

P. Lichtenthal: *Cenni biografici intorno al celebre maestro Wolfgango Amadeo Mozart* (Milan, 1816/*R*)

G.N. von Nissen: *Biographie W.A. Mozart's nach Originalbriefen* (Leipzig, 1828/*R*)

A.D. Oulibicheff: *Nouvelle biographie de Mozart* (Moscow, 1843, Ger. trans., 1847, as *Mozart's Leben und Werke*)

E. Holmes: *The Life of Mozart* (London, 1845, rev. 2/1878/*R* by E. Prout, rev. 3/1912 by E. Newman)

O. Jahn: *W.A. Mozart* (Leipzig, 1856, 2/1867; rev. 3/1889–91, by H. Deiters, 4/1905–7; Eng. trans., 1882) [for later edns see Abert, F 1919–21]

L. Nohl: *Mozart* (Stuttgart, 1863, 2/1877 as *Mozarts Leben*, rev. 3/1906 by P. Sakolowski)

G. Nottebohm: *Mozartiana* (Leipzig, 1880/*R*)

T. de Wyzewa and G. de Saint-Foix: *W.-A. Mozart: sa vie musicale et son oeuvre*, i–ii (Paris, 1912, 2/1936/*R*); iii–v (Paris, 1936–46/*R*) [iii–v by Saint-Foix alone]

A. Schurig: *Wolfgang Amadeus Mozart: sein Leben und sein Werk* (Leipzig, 1913, rev. 2/1923 as *Wolfgang Amadé Mozart: sein Leben, seine Persönlichkeit, sein Werk*)

H. de Curzon: *Mozart* (Paris, 1914, 2/1938)

J.S.J. Kreitmeier: *W.A. Mozart: eine Charakterzeichnung des grossen Meisters nach literarischen Quellen* (Düsseldorf, 1919)

H. Abert: *W.A. Mozart: neu bearbeitete und erweiterte Ausgabe von Otto Jahns 'Mozart'* (Leipzig, 1919–21, 3/1955–66)

L. Schiedermair: *Mozart: sein Leben und seine Werke* (Munich, 1922, enlarged 2/1948)

B. Paumgartner: *Mozart* (Berlin, 1927, enlarged 6/1967, 10/1993)

R. Haas: *Wolfgang Amadeus Mozart* (Potsdam, 1933, 2/1950)

E.F. Schmid: *Wolfgang Amadeus Mozart* (Lübeck, 1934, enlarged 3/1955)

E. Blom: *Mozart* (London, 1935, rev. 6/1974 by J. Westrup)

A. Einstein: *Mozart: his Character, his Work* (New York, 1945; Ger. orig., Stockholm, 1947, 3/1968)

E. Schenk: *Wolfgang Amadeus Mozart: eine Biographie* (Zürich, 1955, rev. 2/1975 as *Mozart: sein Leben, seine Welt*; Eng. trans., abridged, 1960 as *Mozart and his Times*)

J. and B. Massin: *Wolfgang Amadeus Mozart* (Paris, 1959, 3/1990)

A.H. King: *Mozart: a Biography with a Survey of Books, Editions & Recordings* (London, 1970)

M. Levey *The Life and Death of Mozart* (London, 1971, rev. 2/1988)

A. Hutchings *Mozart: the Man, the Musician* (London, 1976)

W. Hildesheimer *Mozart* (Frankfurt, 1977/*R*; Eng. trans., 1982)

I. Keys *Mozart: his Music in his Life* (London, 1980)

S. Sadie: *The New Grove Mozart* (London, 1982)

V. Braunbehrens: *Mozart in Wien* (Munich, 1986; in Eng. trans., 1990)

P.J. Davies: *Mozart in Person* (New York, 1989)

K. Küster: *Mozart: eine musikalische Biographie* (Stuttgart, 1990; Eng. trans., 1996)

N. Elias: *Mozart: zur Sociologie eines Genies*, ed. M. Schröter (Frankfurt, 1991; Eng. trans., 1993, as *Mozart: Portrait of a Genius*)

G. Knepler: *Wolfgang Amadé Mozart: Annäherungen* (Berlin, 1991; Eng. trans., 1994)

M. Solomon: *Mozart: a Life* (New York, 1995)
R. Halliwell: *The Mozart Family: Four Lives in a Social Context* (Oxford, 1998)
D. Schroder: *Mozart in Revolt* (London, 1999)
M. Head: 'Myths of a Sinful Father: Maynard Solomon's "Mozart" ', *ML*, lxxx (1999), 74–85

G: Life: particular aspects and episodes

F. Rochlitz: 'Verbürgte Anekdoten aus Wolfgang Gottlieb Mozarts Leben: ein Beitrag zur richtigeren Kenntnis dieses Mannes, als Mensch und Künstler', *AMZ*, i (1798–9), 17–24, 49–55, 81–6, 113–17, 145–52, 177–80, 289–91, 480–87, 854–6; iii (1800–01), 450–52, 493–7, 590–96
L. Da Ponte: *Memorie* (New York, 1823–7, 2/1829–30; Eng. trans., 1929/*R*)
E. Mörike: *Mozart auf der Reise nach Prag* (Stuttgart, 1856; Eng. trans., 1957) [novel]
C.F. Pohl: *Mozart und Haydn in London* (Vienna, 1867/*R*)
A.J. Hammerle: *Mozart und einige Zeitgenossen* (Salzburg, 1877)
K. Prieger: *Urtheile bedeutender Dichter, Philosopher und Musiker über Mozart* (Wiesbaden, 2/1886)
R. Procházka: *Mozart in Prag* (Prague, 1892; enlarged 2/1938 by P. Nettl as *Mozart in Böhmen*)
E.K. Blümml: *Aus Mozarts Freundes und Familien Kreis* (Vienna, 1923)
O.E. Deutsch: *Mozart und die Wiener Logen: zur Geschichte seiner Freimaurer-Kompositionen* (Vienna, 1932)
H.A. Thies: *Mozart und München: ein Gedenkbuch* (Munich, 1941)
E.F. Schmid: 'Mozart und das geistliche Augsburg', in Deininger, B 1942–3, 40–202
E. Schenk: 'Neues zu Mozarts erster Italienreise: Mozart in Verona', *Neues Mozart-Jb 1943*, 22–44
I. Hoesli: *Wolfgang Amadeus Mozart: Briefstil eines Musikgenies* (Zürich, 1948)
A.B. Gottron: *Mozart und Mainz* (Mainz, 1951)
L. Caflisch and M. Fehr: *Der junge Mozart in Zürich: ein Beitrag zur Mozart-Biographie auf Grund bisher unbekannter Dokumente* (Zürich, 1952)
N. Medici di Marignano and R. Hughes, eds.: *A Mozart Pilgrimage: Being the Travel Diaries of Vincent and Mary Novello in the Year 1829* (London, 1955/*R*)
A. Ostoja: *Mozart e l'Italia* (Bologna, 1955)
E. Schenk: 'Mozart in Mantua', *SMw*, xxii (1955), 1–29
G. Barblan and A. Della Corte, eds.: *Mozart in Italia* (Milan, 1956)
O.E. Deutsch: 'Phantasiestücke aus der Mozart-Biographie', *MJb 1956*, 46–50
L.E. Staehelin: 'Neues zu Mozarts Aufenthalten in Lyon, Genf und Bern', *SMz*, xcvi (1956), 46–8
O.E. Deutsch: 'Aus Schiedenhofens Tagebuch', *MJb 1957*, 15–24
P. Nettl: *Mozart and Masonry* (New York, 1957/*R*)
A. Greither: *Wolfgang Amadé Mozart: seine Lebensgeschichte an Briefen und Dokumenten dargestellt* (Heidelberg, 1958)
W. Hummel, ed.: *Nannerl Mozarts Tagebuchblätter, mit Eintragungen ihres Bruders Wolfgang Amadeus* (Salzburg, 1958) [see also K.P. Pfannhauser, *MISM*, viii/1–2 (1959), 11–17]

E. Winternitz: 'Gnagflow Trazom: an Essay on Mozart's Script, Pastimes, and Nonsense Letters', *JAMS*, xi (1958), 200–216

H.F. Deininger and J. Herz: 'Beiträge zur Genealogie der ältesten schwäbischen Vorfahren W.A. Mozarts', in Deininger, B 1962, 1–76

O.E. Deutsch: 'Mozart in Zinzendorfs Tagebüchern', *SMz*, cii (1962), 211–18

H.W. Hamann: 'Mozarts Schülerkreis', *MJb 1962–3*, 115–39; suppl. by C. Bär, *Acta Mozartiana*, xi (1964), 58–64

O.E. Deutsch: 'Die Legende von Mozarts Vergiftung', *MJb 1964*, 7–18 [with discussion by C. Bär]

L. Wegele: 'Die Mozart: neue Forschungen, zur Ahnengeschichte Wolfgang Amadeus Mozarts', *Acta mozartiana*, xi (1964), 18–24; also in *MISM*, xii/3–4 (1964), 1–6

W. Lievense: *De familie Mozart op bezowk in Nederland: een reisverslag* (Hilversum, 1965)

C. Bär: *Mozart: Krankheit, Tod, Begräbnis*, Schriftenreihe der Internationalen Stiftung Mozarteum, i (Kassel, 1966, 2/1972)

A.R. Mohr: *Das Frankfurter Mozart-Buch* (Frankfurt, 1968)

L.E. Staehelin: *Die Reise der Familie Mozart durch die Schweiz* (Berne, 1968)

A. Greither: *Die sieben grossen Opern Mozarts: mit ein Pathographie Mozarts* (Heidelberg, 2/1970, 3/1977) [orig. pubd 1956 without 'Pathographie']

E. Hintermaier: *Die Salzburger Hofkapelle von 1700 bis 1806: Organisation und Personal* (diss., U. of Salzburg, 1972)

H. Schuler: *Wolfgang Amadeus Mozarts Vorfahren und Gesamtverwandtschaft* (Essen, 1974, rev. 2/1980 as *Wolfgang Amadeus Mozart: Vorfahren u. Verwandte*)

J.H. Eibl: 'Die Mozarts und der Erzbischof', *ÖMz*, xxx (1975), 329–41

M.H. Schmid: *Mozart und die Salzburger Tradition* (Tutzing, 1976)

K. Thomson: 'Mozart and Freemasonry', *ML*, lvi (1976), 25–46

K. Thomson: *The Masonic Thread in Mozart* (London, 1977)

C. Bär: '"Er war ... kein guter Wirth": eine Studie über Mozarts Verhältnis zum Geld', *Acta mozartiana*, xxv (1978), 30–53

E. Hintermaier: *Die Salzburger Hofkapelle von 1700 bis 1806: Organisation und Personal* (diss., U. of Salzburg, 1972)

O. Biba: 'Grundzüge des Konzertwesens in Wien zu Mozarts Zeit', *Mozart und seine Umwelt: Salzburg 1976* [*MJb 1978–9*], 132–43

M. Brown: 'Mozart and After: the Revolution in Musical Consciousness', *Critical Inquiry*, vii (1980–81), 689–706

R. Angermüller: *W.A. Mozarts musikalische Umwelt in Paris (1778): eine Dokumentation* (Munich, 1982)

H.C.R. Landon: *Mozart and the Masons* (London, 1982, 2/1991)

P. Autexier: *Mozart et Liszt sub rosa* (Poitiers, 1984)

P. Davies: 'Mozart's Illnesses and Death', *MT*, cxxv (1984), 437–42, 554–62

A. Steptoe: 'Mozart and Poverty: a Re-Examination of the Evidence', *MT*, cxxv (1984), 196–201

D. Beales: *Joseph II*, i: *In the Shadow of Maria Theresa 1741–1780* (Cambridge, 1987)

H. Dopsch and H. Spatzenegger, eds.: *Geschichte Salzburgs: Stadt und Land*, ii/1 (Salzburg, 1988)

H.-J. Irmen: *Mozart: Mitglied geheimer Gesellschaften* (Mechernich, 1988, 2/1991)

H.C.R. Landon: *1791, Mozart's Last Year* (London, 1988)

P.J. Davies: *Mozart in Person: his Character and Health* (New York, 1989)
C. Eisen: 'Salzburg under Church Rule', *Man & Music: the Classical Era*, ed. N. Zaslaw (London, 1989), 166–87
K. Komlós: 'Mozart and Clementi: a Piano Competition and its Interpretation', *Historical Performance*, ii (1989), 3–9; Hung. orig. in *Muzsika*, xxx/11 (1987), 19–24
H.C.R. Landon: *Mozart: the Golden Years, 1781–1791* (London, 1989)
J. Moore: 'Mozart in the Market-Place', *JRMA*, cxiv (1989), 18–42
M.S. Morrow: *Concert Life in Haydn's Vienna: Aspects of a Developing Musical and Social Institution* (Stuyvesant, NY, 1989)
W. Brauneis: '"... wegen schuldigen 1435 f 32 xr": neuer Archivfund zur Finanzmisere Mozarts im November 1791', *MISM*, xxxix (1991), 159–64
D. Edge: 'Mozart's Fee for Così fan tutte', *JRMA*, cxvi (1991), 211–35
H.C.R. Landon: *Mozart and Vienna* (London, 1991)
W. Stafford: *Mozart's Death: a Corrective Survey of the Legends* (London, 1991)
H. Strebel: *Der Freimaurer Wolfgang Amadé Mozart* (Stäfa, 1991)
H. Schuler: *Mozart und die Freimaurerei: Daten, Fakten, Biographien* (Wilhelmshaven, 1992)
Dalhousie Review, lxxii/2 (1993) [issue on Mozart and contemporary medicine]
P. Clive: *Mozart and his Circle: a Biographical Dictionary* (London, 1993)
G. Fischer-Colbrie: 'Die Mitgliederliste der Freimaurerloge "Zur gekrönten Hoffnung" aus Mozarts Sterbejahr aufgefunden', *MISM*, xli/3–4 (1993), 35–47
S. McVeigh: *Concert Life in London from Mozart to Haydn* (Cambridge, 1993)
R. Angermüller and G. Geffray *Delitiae Italiae: Mozarts Reisen in Italien* (Bad Honnef, 1994)
B.C. Clarke: 'Albert von Mölk: Mozart Myth-Maker? Study of an 18th-Century Correspondence', *MJb 1995*, 155–91
I. Woodfield: 'New Light on the Mozarts' London Visit: a Private Concert with Manzuoli', *ML*, lxxvi (1995), 187–208
D. Beales: 'Court, Government and Society in Mozart's Vienna', *Wolfgang Amadè Mozart: Essays on his Life and Music*, ed. S. Sadie (Oxford, 1996), 3–20
D. Edge: 'Mozart's Reception in Vienna, 1787–91', *Wolfgang Amadè Mozart: Essays on his Life and Music*, ed. S. Sadie (Oxford, 1996), 66–120
N. Zaslaw: 'The Breitkopf Firm's Relations with Leopold and Wolfgang Mozart', *Bach Perspectives*, ii (1996), 85–103
R.M. Ridgewell: *Mozart and the Artaria Publishing House: Studies in the Inventory Ledgers, 1784–1793* (diss., U. of London, 1999)

H: Works: style, influences, particular aspects

A. Heuss: 'Das dämonische Element in Mozarts Werken', *ZIMG*, vii (1905–6), 175–86
G. Schünemann, ed.: *Mozart als achtjähriger Komponist: ein Notenbuch Wolfgangs* (Leipzig, 1909)
R. Lach: *W.A. Mozart als Theoretiker* (Vienna, 1918)
F. Torrefranca: 'Le origini dello stile Mozartiano', *RMI*, xxviii (1921), 263–308; xxxiii (1926), 321–42, 505–29; xxxiv (1927), 1–33, 169–89, 493–511; xxxvi (1929), 373–407
W. Lüthy: *Mozart und die Tonartencharakteristik* (Strasbourg, 1931/*R*)
D.F. Tovey: *Essays in Musical Analysis* (London, 1935–9), i, iii, vi [incl. essays on orch works]

C. Thieme: *Der Klangstil des Mozartorchesters* (Leipzig, 1936)

D.F. Tovey: *Essays in Musical Analysis: Chamber Music*, ed. H.J. Foss (London, 1944/*R*)

A. Einstein: 'Mozart's Choice of Keys', *MQ*, xxvii (1941), 415–21; repr. in Einstein, F 1945, 175–81

A.H. King: 'Mozart's Counterpoint: its Growth and Significance', *ML*, xxvi (1945), 12–20; repr. in King, B 1955, 3/1970/*R*, 164–79

D. Bartha: 'Mozart et le folklore musical de l'Europe centrale', *Les influences étrangères dans l'oeuvre de W.A. Mozart: Paris 1956* (Paris, 1958), 157–81

I.M. Bruce: 'A Note on Mozart's Bar-Rhythms', *MR*, xvii (1956), 35–47

H.T. David: 'Mozartean Modulations', *MQ*, xlii (1956), 193–212; repr. in Lang, B 1963, 56–72

H. Engel: 'Mozarts Instrumentation', *MJb 1956*, 51–74

E.E. Lowinsky: 'On Mozart's Rhythm', *MQ*, xlii (1956), 162–86; repr. in Lang, B 1963, 31–55

E.F. Schmid: 'Mozart and Haydn', *MQ*, xlii (1956), 145–61

B. Szabolcsi: 'Die "Exotismen" Mozarts', *Leben und Werk W.A. Mozarts: Prague 1956* (Prague, ?1958), 181–8; Eng. trans., *ML*, xxxvii (1956), 323–32

L.F. Tagliavini: 'L'opéra italien du jeune Mozart', *Les influences étrangères dans l'oeuvre de W.A. Mozart: Paris 1956* (Paris, 1958), 125–56

E.J. Dent and E. Valentin: *Der früheste Mozart* (Munich, 1956) [in Ger. and Eng.]

E. Valentin, ed.: *L. Mozart: Nannerls Notenbuch, 1759* (Munich, 1956)

W. Siegmund-Schultze: *Mozarts Melodik und Stil* (Leipzig, 1957)

H. Engel: 'Haydn, Mozart und die Klassik', *MJb 1959*, 46–79

G. Massenkeil: *Untersuchungen zum Problem der Symmetrie in der Instrumentalmusik W.A. Mozarts* (Wiesbaden, 1962)

H. Engel: 'Nochmals: thematische Satzverbindungen und Mozart', *MJb 1962–3*, 14–23

I.R. Eisley: 'Mozart and Counterpoint: Development and Synthesis', *MR*, xxiv (1963), 23–9

S.G. Davis: 'Harmonic Rhythm in Mozart's Sonata Form', *MR*, xxvii (1966), 25–43

W. Kirkendale: *Fuge und Fugato in der Kammermusik des Rokoko und der Klassik* (Tutzing, 1966), 184–215; Eng. trans., enlarged (Durham, NC, 1979), 152–81

H. Beck: 'Harmonisch-melodische Modelle bei Mozart', *MJb 1967*, 90–99

M.S. Cole: 'The Rondo Finale: Evidence for the Mozart–Haydn Exchange?', *MJb 1968–70*, 242–56

M. Flothuis: *Mozarts Bearbeitungen eigener und fremder Werke*, Schriftenreihe der Internationalen Stiftung Mozarteum, ii (Salzburg and Kassel, 1969)

K.J. Marx: *Zur Einheit der zyklischen Form bei Mozart* (Stuttgart, 1971)

C. Rosen: *The Classical Style: Haydn, Mozart, Beethoven* (London and New York, 1971, enlarged 3/1997 with sound disc)

H. Federhofer: 'Mozart als Schüler und Lehrer in der Musiktheorie', *MJb 1971–2*, 89–106

H. Federhofer and others: 'Tonartenplan und Motivstruktur (Leitmotivtechnik?) in Mozarts Musik', *Idomeneo Conference: Salzburg 1973* [*MJb 1973–4*], 82–144 [discussions]

D. Heartz: 'Thomas Attwood's Lessons in Composition with Mozart', *PRMA*, c (1973–4), 175–83

W. Plath and others: 'Typus und Modell in Mozarts Kompositionsweise', *Idomeneo Conference: Salzburg 1973* [*MJb 1973–4*], 145–78 [discussions]
U. Toeplitz: *Die Holzbläser in der Musik Mozarts und ihr Verhältnis zur Tonartwahl* (Baden-Baden, 1978)
C. Rosen: *Sonata Forms* (New York, 1980, 2/1988)
F.K. Grave: '"Rhythmic Harmony" in Mozart', *MR*, xli (1980), 87–102
P.A. Autexier: *Les oeuvres témoins de Mozart* (Paris, 1982)
A.P. Brown: 'Haydn and Mozart's 1773 Stay in Vienna: Weeding a Musicological Garden', *JM*, x (1992), 192–230
H. Krones: 'Barocke Traditionen in der österreichischen Musik des späten 18. und frühen 19. Jahrhunderts', *Alte Musik ... Perspektiven der Aufführungspraxiz: Graz 1992* (Regensburg, 1994), 65–92
E.R. Sisman: *Haydn and the Classical Variation* (Cambridge, MA, 1993)
C. Wolff: 'Vollendet und fragmentarisch: Über Mozarts Schaffen der letzten Lebensjahre', *Jb alte Musik*, ii (1993), 61–87
C. Eisen: 'Mozart e l'Italia: il ruolo di salisburgo', *RIM*, xxx (1995), 51–84
J. Garland: 'Form, Genre, and Style in the Eighteenth-Century Rondo', *Music Theory Spectrum*, xvii (1995), 27–52
D. Heartz: *Haydn, Mozart and the Viennese School, 1740–1780* (New York, 1995)
S.B. Jan: *Aspects of Mozart's Music in G minor: toward the Identification of Common Structural and Compositional Characteristics* (New York, 1995)
H.S. Powers: 'Reading Mozart's Music: Text and Topic, Sense and Syntax', *CMc*, no.57 (1995), 5–44
R. Kamien and N. Wagner: 'Bridge Themes within a Chromaticized Voice Exchange in Mozart Expositions', *Music Theory Spectrum*, xix (1997), 1–12

I: Sacred works

W. Pole: *The Story of Mozart's Requiem* (London, 1879) [first printed in *MT*, xiv (1869–71), 39–41, 71–4, 103–7, 135–7, 167–70, 201–4, 237–43]
K.A. Rosenthal: 'The Salzburg Church Music of Mozart and his Predecessors', *MQ*, xviii (1932), 559–77
K.A. Rosenthal: 'Mozart's Sacramental Litanies and their Forerunners', *MQ*, xxvii (1941), 433–55
K.G. Fellerer: *Mozarts Kirchenmusik* (Salzburg, 1955)
G. Reichert: 'Mozarts "Credo-Messen" und ihre Vorläufer', *MJb 1955*, 117–44
K. Geiringer: 'The Church Music', in Landon and Mitchell, B 1956, 361–76
H. Federhofer: 'Probleme der Echtheitsbestimmung der kleineren kirchenmusikalischen Werke W.A. Mozarts', *MJb 1958*, 97–108; suppl., *MJb 1960–61*, 43–51
K. Pfannhauser: 'Mozarts kirchenmusikalische Studien im Spiegel seiner Zeit und Nachwelt', *KJb*, xliii (1959), 155–98
F. Blume: 'Requiem but no Peace', *MQ*, xlvii (1961), 147–69; repr. in Lang, B 1963, 103–26
O.E. Deutsch: 'Zur Geschichte von Mozarts Requiem', *ÖMz*, xix (1964), 49–60
L. Nowak: 'Das Requiem von W.A. Mozart', *ÖMz*, xx (1965), 395–9
R. Federhofer-Königs: 'Mozarts "Lauretanische Litaneien" KV 109 (74e) und 195 (186d)', *MJb 1967*, 111–20

A. Holschneider: 'C.Ph.E. Bachs Kantate "Auferstehung und Himmelfahrt Jesu" und Mozarts Aufführung des Jahres 1788', *MJb 1968–70*, 264–80
F. Beyer: '"Mozarts Komposition zum Requiem": zur Frage der Ergänzung', *Acta mozartiana*, xviii (1971), 27–33
G. Duda: 'Neues aus der Mozartforschung: Requiem-Begräbnis-Grabfrage', *Acta mozartiana*, xviii (1971), 33–7 [on the Requiem]
C. Rosenthal: 'Der Einfluss der Salzburger Kirchenmusik auf Mozarts kirchenmusikalische Kompositionen', *MJb 1971–2*, 173–81
L. Nowak: 'Wer hat die Instrumentalstimmen in der Kyrie-Fuge des Requiems von W.A. Mozart geschrieben? Ein vorläufiger Bericht', *Idomeneo Conference: Salzburg 1973* [*MJb 1973–4*], 191–201
K.G. Fellerer: 'Liturgische Grundlagen der Kirchenmusik Mozarts', *Festschrift Walter Senn*, ed. E. Egg and E. Fässler (Munich, 1975), 64–74
'Sektion Kirchenmusik', *MJb 1978–9*, 14–29 [4 articles]
R. Maunder: *Mozart's Requiem: on Preparing a New Edition* (Oxford, 1988)
P. Moseley: 'Mozart's Requiem: a Revaluation of the Evidence', *JRMA*, cxiv (1989), 203–37
C. Wolff: *Mozarts Requiem: Geschichte, Musik, Dokumente, Partitur des Fragments* (Munich and Kassel, 1991; Eng. trans., 1994)
R.D. Levin: 'Zu den von Süssmayr komponierten Sätzen des Requiems KV 626', *Mozart Congress: Salzburg 1991* [*MJb 1991*], 475–93
W. Brauneis: '"Dies irae, dies illa – Tag des Zornes, Tag der Klage": Auftrag, Entstehung und Vollendung von Mozarts "Requiem"', *Jb des Vereins für Geschichte der Stadt Wien*, xlvii–xlviii (1991–2), 33–50
R. Münster: 'Die beiden Fassungen der Motette *Exsultate, jubilate* KV 165', *Mozart-Studien*, ii (1993), 119–33
W.-D. Seiffert: 'Wolfgang Amadeus Mozart: Requiem', *Werkanalyse in Beispielen: grosse Chorwerke*, ed. S. Helms and R. Schneider (Kassel, 1994), 72–97
D. Leeson: 'Franz Xaver Süssmayr and the Mozart Requiem: a Computer Analysis of Authorship based on Melodic Affinity', *MJb 1995*, 111–53
M. Schuler: 'Mozarts Requiem in der Tradition gattungsgeschichtlicher Topoi', *Studien zur Musikgeschichte: eine Festschrift für Ludwig Finscher*, ed. A. Laubenthal and K. Kusan-Windweh (Kassel, 1995), 317–27
B.C. Clarke: 'From Little Seeds: what were the Circumstances Surrounding the Inception of Mozart's Requiem and its Aftermath?', *MT*, cxxxvii (1996), 13–17

J: Operas

A.D. Oulibicheff: *Mozarts Opern: kritische Erläuterungen* (Leipzig, 1848) [trans. of part of Oulibicheff, F 1843]
C. Gounod: *Le Don Juan de Mozart* (Paris, 1890/*R*, Eng. trans. of 3rd edn, 1895/*R*)
E. Komorzynski: *Emanuel Schikaneder: ein Beitrag zur Geschichte des deutschen Theaters* (Vienna, 1901, 2/1951)
E.J. Dent: *Mozart's Operas: a Critical Study* (London, 1913, 2/1947/*R*)
A. Lorenz: 'Das Finale in Mozarts Meisteropern', *Die Musik*, xix (1926–7), 621–32
E. Blom: 'The Literary Ancestry of Figaro', *MQ*, xiii (1927), 528–39

P. Stefan: *Die Zauberflöte: Herkunft, Bedeutung, Geheimnis* (Vienna, 1937)
E. Komorzynski: '"Die Zauberflöte": Entstehung und Bedeutung des Kunstwerkes', *Neues Mozart-Jb 1941*, 147–74
H.F. Redlich: 'L'oca del Cairo', *MR*, ii (1941), 122–31
E. Wellesz: 'Don Giovanni and the dramma giocoso', *MR*, iv (1943), 121–6
C. Benn: *Mozart on the Stage* (London, 1946)
S. Levarie: *Mozart's "Le nozze di Figaro": a Critical Analysis* (Chicago, 1952/*R*)
H. Engel: 'Die Finali der Mozartschen Opern', *MJb 1954*, 113–34
Les influences étrangères dans l'oeuvre de W.A. Mozart: Paris 1956 (Paris, 1958)
A. Greither: *Die sieben grossen Opern Mozarts: Versuche über das Verhältnis der Texte zur Musik* (Heidelberg, 1956, enlarged 2/1970, 3/1977)
C. Raeburn: 'Die textlichen Quellen des "Schauspieldirektor" ', *ÖMz*, xiii (1958), 4–10
T. Volek: 'Über den Ursprung von Mozarts Oper "La clemenza di Tito" ', *MJb 1959*, 274–86
C. Bitter: *Wandlungen in den Inszenierungsformen des 'Don Giovanni' von 1787 bis 1928* (Regensburg, 1961)
S. Kunze: 'Mozarts Schauspieldirektor', *MJb 1962–3*, 156–67
F.-H. Neumann: 'Zur Vorgeschichte der Zaide', *MJb 1962–3*, 216–47
B. Brophy: *Mozart the Dramatist: a New View of Mozart, his Operas and his Age* (London, 1964, rev. 1988 as *Mozart the Dramatist: the Value of his Operas to him, to his Age, and to Us*)
C. Floros: 'Das "Programm" in Mozarts Meisterouvertüren', *SMw*, xxvi (1964), 140–86
C. Raeburn: 'Die Entführungsszene aus "Die Entführung aus dem Serail" ', *MJb 1964*, 130–37
A. Rosenberg: *Die Zauberflöte: Geschichte und Deutung* (Munich, 1964, 3/1981)
R. Moberly and C. Raeburn: 'Mozart's "Figaro": the Plan of Act III', *ML*, xlvi (1965), 134–6; repr. in *MJb 1965–6*, 161–3
R. Münster: 'Die verstellte Gärtnerin: neue Quellen zur authentischen Singspielfassung von W.A. Mozarts La finta giardiniera', *Mf*, xviii (1965), 138–60
D.J. Keahey: 'Così fan tutte: Parody or Irony?', *Paul A. Pisk: Essays in his Honor*, ed. J. Glowacki (Austin, 1966), 116–30
A.A. Abert: 'Beiträge zur Motivik von Mozarts Spätopern', *MJb 1967*, 7–11
F. Giegling: 'Zu den Rezitativen von Mozarts Oper "Titus" ', ibid., 121–6
G. Gruber: 'Das Autograph der "Zauberflöte"', ibid., 127–49; *MJb 1968–70*, 99–110
D. Heartz: 'The Genesis of Mozart's "Idomeneo" ', *MJb 1967*, 150–64; repr. in *MQ*, lv (1969), 1–19
K.-H. Köhler: 'Mozarts Kompositionsweise: Beobachtungen am Figaro-Autograph', *MJb 1967*, 31–45
R.B. Moberly: *Three Mozart Operas: Figaro, Don Giovanni, The Magic Flute* (London, 1967)
A.A. Abert: '"La finta giardiniera" und "Zaide" als Quellen für spätere Opern Mozarts', *Musik und Verlag: Karl Vötterle zum 65. Geburtstag*, ed. R. Baum and W. Rehm (Kassel, 1968), 113–22
J. Chailley: *La flûte enchantée, opéra maçonnique: essai d'explication du livret et de la musique* (Paris, 1968, 2/1983; Eng. trans., 1971/*R*, 2/1992 as *The Magic Flute Unveiled: Esoteric Symbolism in Mozart's Masonic Opera*)
F.R. Noske: 'Musical Quotation as a Dramatic Device: the Fourth Act of "Le nozze di Figaro" ', *MQ*, liv (1968), 185–98; repr. in Noske, J 1977

L.F. Tagliavini: 'Quirino Gasparini and Mozart', *New Looks at Italian Opera: Essays in Honor of Donald J. Grout*, ed. W.W. Austin (Ithaca, NY, 1968), 151–71 [on *Mitridate*]
S. Döhring: 'Die Arienformen in Mozarts Opern', *MJb 1968–70*, 66–76
H. Federhofer: 'Die Harmonik als dramatischer Ausdrucksfaktor in Mozarts Meisteropern', ibid., 77–87
F. Giegling: 'Metastasios Oper "La clemenza di Tito" in der Bearbeitung durch Mazzola', ibid., 88–94
K.-H. Köhler: 'Figaro-Miscellen: einige dramaturgische Mitteilungen zur Quellensituation', ibid., 119–31
C.-H. Mahling: 'Typus und Modell in Opern Mozarts', ibid., 145–58
C. Henning: 'Thematic Metamorphoses in Don Giovanni', *MR*, xxx (1969), 22–6
F. Noske: 'Social Tensions in "Le nozze di Figaro"', *ML*, l (1969), 45–62; repr. in Noske, J 1977, 18–38
A.A. Abert: *Die Opern Mozarts* (Wolfenbüttel, 1970); abridged Eng. trans., *NOHM*, vii (1973), 97–171
B. Brophy: '"Figaro" and the Limitations of Music', *ML*, li (1970), 26–36
F.R. Noske: '"Don Giovanni": Musical Affinities and Dramatic Structure', *SMH*, xii (1970), 167–203; repr. in *Theatre Research/Recherches téâtrales*, viii (1973), 60–74 and in Noske, J 1977, 39–75
A. Williamson: 'Who was Sarastro?', *Opera*, xxi (1970), 297–305; see also 695–6
H. Keller: 'Mozart's Wrong Key Signature', *Tempo*, no.98 (1971), 21–7 [on *Così fan tutte*]
H.H. Eggebrecht: *Versuch über die Wiener Klassik: die Tanzszene in Mozarts 'Don Giovanni'* (Wiesbaden, 1972)
S. Kunze: *Don Giovanni vor Mozart: die Tradition der Don-Giovanni-Opern im italienischen Buffo-Theater des 18. Jahrhunderts* (Munich, 1972)
H. Goldschmidt: 'Die Cavatina des Figaro: eine semantische Analyse', *BMw*, xv (1973), 185–207
S. Kunze: 'Über das Verhältnis von musikalisch autonomer Struktur und Textbau in Mozarts Opern', *Idomeneo Conference: Salzburg 1973* [*MJb 1973–4*], 217–32
R.B. Moberly: 'Mozart and his Librettists', *ML*, liv (1973), 161–9
B. Williams: 'Passion and Cynicism: Remarks on "Così fan tutte"', *MT*, cxiv (1973), 361–4
D. Heartz: 'Raaff's last Aria: a Mozartian Idyll in the Spirit of Hasse', *MQ*, lx (1974), 517–43 [from *Idomeneo*]
D. Heartz: 'Tonality and Motif in Idomeneo', *MT*, cxv (1974), 382–6
K. Hortschansky: 'Mozarts "Ascanio in Alba" und der Typus der Serenata', *Mozart und Italien: Rome 1974* [*AnMc*, no.18 (1978)], 148–59
H. Lühning: 'Zur Entstehungsgeschichte von Mozarts "Titus"', *Mf*, xxvii (1974), 300–318; see also xxviii (1975), 75–81, 311–14
R.B. Moberly: 'The Influence of French Classical Drama on Mozart's "La clemenza di Tito"', *ML*, lv (1974), 286–98
G. Gruber: 'Bedeutung und Spontaneität in Mozarts "Zauberflöte"', *Festschrift Walter Senn*, ed. E. Egg and E. Fässler (Munich, 1975), 118–30
D. Koenigsberger: 'A New Metaphor for Mozart's *Magic Flute*', *European Studies Review*, v (1975), 229–75
H.L. Scheel: '"Le mariage de Figaro" von Beaumarchais und das Libretto der "Nozze di Figaro" von Lorenzo Da Ponte', *Mf*, xxviii (1975), 156–73

C. Gianturco: *Le opere del giovane Mozart* (Pisa, 1976, 2/1978)
D. Heartz: 'Mozart and his Italian Contemporaries: "La clemenza di Tito"', *Mozart und seine Umwelt: Salzburg 1976* [*MJb 1978–9*], 275–93
R. Angermüller: 'Wer war der Librettist von Mozarts "La finta giardiniera"?', *MJb 1976–7*, 1–20
W. Mann: *The Operas of Mozart* (London, 1977)
F. Noske: *The Signifier and the Signified: Studies in the Operas of Mozart and Verdi* (The Hague, 1977)
D. Heartz: 'Mozart, his Father and "Idomeneo"', *MT*, cxix (1978), 228–31
D. Heartz: 'Mozart's Overture to Titus as Dramatic Argument', *MQ*, lxiv (1978), 29–49
C. Osborne: *The Complete Operas of Mozart* (London, 1978)
S. Vill, ed.: *Così fan tutte: Beiträge zur Wirkungsgeschichte von Mozarts Oper* (Bayreuth, 1978)
C. Floros: *Mozart-Studien*, i: *Zu Mozarts Sinfonik, Opern- und Kirchenmusik* (Wiesbaden, 1979)
J. Parakilas: *Mozart's 'Tito' and the Music of Rhetorical Strategy* (diss., Cornell U., 1979)
D. Heartz: 'The Great Quartet in Mozart's *Idomeneo*', *Music Forum*, v (1980), 233–56
K. Pahlen, ed.: *Wolfgang Amadeus Mozart: Die Entführung aus dem Serail* (Munich, 1980, 4/1997) [text and commentary]
'Mozart und die Oper seiner Zeit', *HJbMw*, v (1981), 115–266 [incl. articles by S. Kunze, C. Floros, G. Gruber, F. Lippmann, C.-H. Mahling, H. Lühning]
J. Rushton: *W.A. Mozart: Don Giovanni* (Cambridge, 1981)
W.J. Allanbrook: 'Pro Marcellina: the Shape of "Figaro", Act IV', *ML*, lxiii (1982), 69–84
R. Münster: 'Neues zum Münchner *Idomeneo* 1781', *Acta mozartiana*, xxxix (1982), 10–20
S. Puntscher Riekmann *Mozart, ein bürgerlicher Künstler: Studien zu den Libretti 'Le nozze di Figaro', 'Don Giovanni' und 'Così fan tutte'* (Vienna, 1982)
L'avant-scène opéra, no.54 (1983) [*Mitridate* issue]
W.J. Allanbrook: *Rhythmic Gesture in Mozart: Le nozze di Figaro & Don Giovanni* (Chicago, 1983)
S. Kunze: *Mozarts Opern* (Stuttgart, 1984)
J. Platoff: *Music and Drama in the Opera Buffa Finale: Mozart and his Contemporaries in Vienna, 1781–1790* (diss., U. of Pennsylvania, 1984)
C. Wolff: '"O ew'ge Nacht! Wann wirst du schwinden?" Zum Verständnis der Sprecherszene im ersten Finale von Mozarts Zauberflöte', *Analysen: Beiträge zu einer Problemgeschichte des Komponierens: Festschrift für Hans Heinrich Eggebrecht*, ed. W. Breig, R. Brinkmann and E. Buddle (Wiesbaden, 1984), 234–47
W. Allanbrook: 'Mozart's Happy Endings: a New Look at the "Convention" of the "lieto fine"', *MJb 1984–5*, 1–5
M.P. McClymonds: 'Mozart's "La clemenza di Tito" and opera seria in Florence as a Reflection of Leopold II's Musical Taste', ibid., 61–70
I. Nagel: *Autonomie und Gnade: über Mozarts Opern* (Munich, 1985, 3/1988; Eng. trans., 1991)
S. Henze-Döhring: *Opera seria, opera buffa, und Mozarts Don Giovanni*, AnMc, no.24 (1986)
B.A. Brown: 'Beaumarchais, Mozart and the Vaudeville: Two Examples from "The Marriage of Figaro"', *MT*, cxxvii (1986), 261–5

M. Freyhan: 'Toward the Original Text of Mozart's *Die Zauberflöte*', *JAMS*, xxxix (1986), 255–80
T. Bauman: *W.A. Mozart: Die Entführung aus dem Serail* (Cambridge, 1987)
T. Carter: *W.A. Mozart: Le nozze di Figaro* (Cambridge, 1987)
J.A. Rice: *Emperor and Impresario: Leopold II and the Transformation of Viennese Musical Theater, 1790–92* (diss., U. of California, Berkeley, 1987)
J. Webster: 'To Understand Verdi and Wagner we must Understand Mozart', *19CM*, xi (1987–8), 175–93
W.J. Allanbrook: '*Opera seria* Borrowings in *Le nozze de Figaro*: the Count's "Vedrò mentr'io"', *Studies in the History of Music*, ii (1988), 83–96
D. Neville: 'Cartesian Principles in Mozart's *La clemenza di Tito*', ibid., 97–123
J. Platoff: 'Writing about Influences: *Idomeneo*, a Case Study', *Explorations in Music, the Arts, and Ideas: Essays in Honor of Leonard B. Meyer*, ed. E. Narmour and R.A. Solie (Stuyvesant, NY, 1988), 43–65
A. Steptoe: *The Mozart – Da Ponte Operas* (Oxford, 1988)
A. Tyson: 'The 1786 Prague Version of Mozart's "Le nozze di Figaro"', *ML*, lxix (1988), 321–33
R. Farnsworth: 'Così fan tutte as Parody and Burlesque', *OQ*, vi/2 (1988–9), 50–68
P. Gallarati: 'Music and Masks in Lorenzo Da Ponte's Mozartian Librettos', *COJ*, i (1989), 225–47
V. Mattern: *Das dramma giocoso La finta giardiniera: ein Vergleich der Vertonungen von Pasquale Anfossi und Wolfgang Amadeus Mozart* (Laaber, 1989)
J. Platoff: 'Musical and Dramatic Structure in the Opera Buffa Finale', *JM*, vii (1989), 191–230
L'avant-scène opéra, no.131 (1990) [*Il re pastore* and *Il sogno di Scipione* issue]
C. Abbate and R. Parker: 'Dismembering Mozart', *COJ*, ii (1990), 187–95
D. Heartz and T. Bauman: *Mozart's Operas* (Berkeley, 1990) [incl. T. Bauman: 'At the North Gate: Instrumental Music in *Die Zauberflöte*', 277–97]
D. Link: 'The Viennese Operatic Canon and Mozart's *Così fan tutte*', *MISM*, xxxviii (1990), 111–21
J. Platoff: 'The Buffa Aria in Mozart's Vienna', *COJ*, ii (1990), 99–120
L.L. Tyler: '*Bastien und Bastienne*: the Libretto, its Derivation, and Mozart's Text-Setting', *JM*, viii (1990), 520–52
J. Webster: 'Mozart's Operas and the Myth of Musical Unity', *COJ*, ii (1990), 197–218
P. Branscombe: *W.A. Mozart: Die Zauberflöte* (Cambridge, 1991)
C. Ford: *Così? Sexual Politics in Mozart's Operas* (Manchester, 1991)
J. Platoff: 'Tonal Organization in the "Buffo" Finales and the Act II Finale of *Le nozze di Figaro*', *ML*, lxxii (1991), 387–403
J.A. Rice: *W.A. Mozart: La clemenza di Tito* (Cambridge, 1991)
L. Tyler: '"Zaide" in the Development of Mozart's Operatic Language', *ML*, lxxii (1991), 214–35
J. Webster: 'The Analysis of Mozart's Arias', in Eisen, B 1991, 101–99
R.R. Subotnik: 'Whose *Magic Flute*? Intimations of Reality at the Gates of Enlightenment', *19CM*, xv (1991–2), 132–50
D.J. Buch: 'Fairy-Tale Literature and *Die Zauberflöte*', *AcM*, lxiv (1992), 30–49

J.A. Eckelmeyer: *The Cultural Context of Mozart's Magic Flute: Social, Aesthetic, Philosophical* (Lewiston, NY, 1992)
N. Till: *Mozart and the Enlightenment: Truth, Virtue and Beauty in Mozart's Operas* (London, 1992)
W. Allanbrook: 'Human Nature in the Unnatural garden: Figaro as Pastoral', *CMc*, no.51 (1993), 82–93
W. Brauneis: 'Das Frontispiz im Alberti-Libretto von 1791 als Schlüssel zu Mozarts Zauberflöte', *MISM*, xli/3–4 (1993), 49–59
S. Durante: *Mozart and the Idea of vera opera: a Study of La clemenza di Tito* (diss., Harvard U., 1993)
P. Gallarti: *La forza della parole: Mozart drammaturgo* (Turin, 1993)
E.J. Goehring: *The Comic Vision of 'Così fan tutte': Literary and Operatic Traditions* (diss., Columbia U., 1993)
J. Rushton: *W.A. Mozart: Idomeneo* (Cambridge, 1993)
G.A. Wheelock: 'Schwarze Gredel and the Engendered Minor Mode in Mozart's Operas', *Musicology and Difference: Gender and Sexuality in Music Scholarship*, ed. R.A. Solie (Berkeley, 1993)
H.J. Wignall: 'Guglielmo d'Ettore: Mozart's First Mitridate', *OQ*, x/3 (1993–4), 93–112
S.G. Burnham: 'Mozart's *felix culpa*: *Così fan tutte* and the Irony of Beauty', *MQ*, lxxviii (1994), 77–98
J. Waldoff: 'The Music of Recognition: Operatic Enlightenment in The Magic Flute', *ML*, lxxv (1994), 214–35
J. Rice: 'Mozart and his Singers: the Case of Maria Marchetti Fantozzi, the First Vitellia', *OQ*, xi/4 (1994–5), 31–52
B.A. Brown: *W.A. Mozart: Così fan tutte* (Cambridge, 1995)
J.E. Everson: 'Of Beaks and Geese: Mozart, Varesco, and Francesco Cieco', *ML*, lxxvi (1995), 369–83
E. Goehr: 'Despina, Cupid and the Pastoral Mode of *Così fan tutte*', *COJ*, vii (1995), 107–33
D. Heartz: 'Mozart and Da Ponte', *MQ*, lxxix (1995), 700–18
U. Konrad: '"... mithin liess ich meinen gedanken freyen lauf": erste Überlegungen und Thesen zu den "Fassungen" von W.A. Mozarts Die Entführung aus dem Serail KV 384', *Opernkomposition als Prozess: Bochum 1995*, 47–64
J.A. Rice: 'Leopold II, Mozart, and the Return to a Golden Age', *Opera and the Enlightenment*, ed. T. Bauman and M. McClymonds (Cambridge, 1995), 271–96
J. Waldoff: *The Music of Recognition in Mozart's Operas* (diss., Cornell U., 1995)
P.L. Gidwitz: 'Mozart's Fiordiligi: Adriana Ferrarese dal Bene', *COJ*, viii (1996), 199–214
R. Peiretti: '"Vado incontro al fato estremo": eine bisher fälschlich Mozart zugeschriebene Arie der Oper "Mitridate, re di Ponto"', *MISM*, xliv/3–4 (1996), 40–41
R.J. Rabin: *Mozart, Da Ponte, and the Dramaturgy of Opera Buffa* (diss., Cornell U., 1996)
M. Hunter: 'Rousseau, the Countess, and the Female Domain', in Eisen, B 1997, 1–26
M. Hunter and J. Webster, eds.: *Opera Buffa in Mozart's Vienna* (Cambridge, 1997)
J. Platoff: 'Tonal Organization in the *opera buffa* of Mozart's Time', in Eisen, B 1997, 139–74
M. Hunter: *The Culture of Opera Buffa in Mozart's Vienna: a Poetics of Entertainment* (Princeton, NJ, 1999)

K: Arias, songs and other vocal music

M.J.E. Brown: 'Mozart's Songs for Voice and Piano', *MR*, xvii (1956), 19–28

A. Orel: 'Mozarts Beitrag zum deutschen Sprechtheater: die Musik zu Geblers "Thamos" ', *Acta mozartiana*, iv (1957), 43–53, 74–81

H. Engel: 'Hasses Ruggiero und Mozarts Festspiel Ascanio', *MJb 1960–61*, 29–42

C.B. Oldman: 'Mozart's Scena for Tenducci', *ML*, xlii (1961), 44–52

S. Kunze: 'Die Vertonungen der Arie "Non sò d'onde viene" von J.Chr. Bach und von W.A. Mozart', *AnMc*, no.2 (1965), 85–111

A. Dunning: 'Mozarts Kanons', *MJb 1971–2*, 227–40

S. Dahms: 'Mozarts festa teatrale "Ascanio in Alba" ', *ÖMz*, xxxi (1976), 15–24

E.A. Ballin: *Das Wort-Ton-Verhältnis in den klavierbegleiteten Liedern Mozarts* (Kassel, 1984)

P. Autexier: 'La musique maçonnique', *Dix-huitième siècle*, xix (1987), 97–104

J. Page and D. Edge: 'A Newly Uncovered Autograph Sketch for Mozart's Al desio di chi t'adora K.577', *MT*, cxxxii (1991), 601–6

P.A. Autexier: 'Cinq Lieder inconnus de W.A. Mozart', *International Journal of Musicology*, i (1992), 67–79

D. Edge: 'A Newly Discovered Autograph Source for Mozart's Aria K. 365a (Anh. 11a)', *MJb 1996*, 177–96

D.J. Buch: 'Mozart and the Theater auf der Wieden: New Attributions and Perspectives', *COJ*, ix (1997), 195–232

L: Symphonies, serenades, etc.

D. Schultz: *Mozarts Jugendsinfonien* (Leipzig, 1900)

S. Sechter: *Das Finale der Jupiter-Symphonie (C dur) von W.A. Mozart*, ed. F. Eckstein (Vienna, 1923)

H. Schenker: 'Mozart: Sinfonie g-moll', *Das Meisterwerk in der Musik*, ii (Munich, 1926/*R*), 105–57; (Eng. trans. Cambridge, 1996), 59–96

A.E.F. Dickinson: *A Study of Mozart's Last Three Symphonies* (London, 1927/*R*, 2/1939)

G. de Saint-Foix: *Les symphonies de Mozart* (Paris, 1932; Eng. trans., 1947/*R*)

N. Broder: 'The Wind-Instruments in Mozart's Symphonies', *MQ*, xix (1933), 238–59

G. de Saint-Foix: 'La jeunesse de Mozart: 1771: les diverses orientations de la symphonie', *MJb 1950*, 14–23, 116–26

H. Engel: 'Über Mozarts Jugendsinfonien', *MJb 1951*, 22–33

G. Hausswald: *Mozarts Serenaden: ein Beitrag zur Stilkritik des 18. Jahrhunderts* (Leipzig, 1951/*R*)

H. Engel: 'Der Tanz in Mozarts Kompositionen', *MJb 1952*, 29–39

J.N. David: *Die Jupiter-Symphonie: eine Studie über die thematisch-melodischen Zusammenhänge* (Göttingen, 1953, 4/1960)

H. Beck: 'Zur Entstehungsgeschichte von Mozarts D-Dur-Sinfonie, KV. 297: Probleme der Kompositionstechnik und Formentwicklung in Mozarts Instrumentalmusik', *MJb 1955*, 95–112

H. Keller: 'KV.503: the Unity of Contrasting Themes and Movements', *MR*, xvii (1956), 48–58, 120–29

E.F. Schmid: 'Zur Entstehungszeit von Mozarts italienischen Sinfonien', *MJB 1958*, 71–8

A.A. Abert: 'Stilistischer Befund und Quellenlage: zu Mozarts Lambacher Sinfonie KV Anh.221 = 45a', *Festschrift Hans Engel*, ed. H. Heussner (Kassel, 1964), 43–56

W. Plath and others: 'Echtheitsfragen', *MJb 1971–2*, 17–67 [with discussions of KAnh.9/Anh.C14.01 and K84/73*q*]

P. Benary: 'Metrum bei Mozart: zur metrischen Analyse seiner letzten drei Sinfonien', *SMz*, cxiv (1974), 201–5

S. Wollenberg: 'The Jupiter Theme: New Light on its Creation', *MT*, cxvi (1975), 781–3

L. Meyer: 'Grammatical Simplicity and Relational Richness: the Trio of Mozart's G minor Symphony', *Critical Inquiry*, ii (1975–6), 693–761

R. Dearling: *The Music of Wolfgang Amadeus Mozart: the Symphonies* (London and Rutherford, NJ, 1982)

R. Münster: 'Neue Funde zu Mozarts symphonischen Jugendwerk', *MISM*, xxx.1–2 (1982), 2–11

E. Smith: *Mozart Serenades, Divertimenti and Dances* (London, 1982)

Die Sinfonie KV 16a 'del Sigr. Mozart': Odense 1984

R.R. Subotnik: 'Evidence of a Critical World View in Mozart's Last Three Symphonies', *Music and Civilisation: Essays in Honor of Paul Henry Lang*, ed. E. Strainchamps, M.R. Maniates and C. Hatch (New York, 1984), 29–43

P. Autexier: 'Wann wurde di Maurerische Trauermusik uraufgeführt?', *MJb 1984*, 56–8

N. Zaslaw and C. Eisen: 'Signor Mozart's Symphony in A minor K.Anh.220 = 16a', *JM*, iv (1985–6), 191–206

C. Eisen: 'New Light on Mozart's "Linz" Symphony, K.425', *JRMA*, cxiii (1988), 81–96

L. Treitler: 'Mozart and the Idea of Absolute Music', *Das musikalische Kunstwerk: Festschrift Carl Dahlhaus*, ed. H. Danuser and others (Laaber, 1988), 413–40; repr. in *Music and the Historical Imagination* (Cambridge, MA, 1989), 176–214

N. Zaslaw: *Mozart's Symphonies: Context, Performance Practice, Reception* (Oxford, 1989)

C. Eisen: 'Problems of Authenticity Among Mozart's Early Symphonies: the Examples of K.Anh.220 (16a) and 76 (42a)', *ML*, lxx (1989), 505–16

L. Dreyfus: 'The Hermeneutics of Lament: a Neglected Paradigm in a Mozartian Trauermusik', *Music Analysis*, x (1991), 329–43

D. Blazin: 'The Two Versions of Mozart's Divertimento K.113', *ML*, lxxiii (1992), 32–47

W. Gersthofer: *Mozarts frühe Sinfonien (bis 1772): Aspekte frühklassischer Sinfonik* (Kassel, 1993)

S. Sechter: 'Analysis of the Finale of Mozart's Symphony no.41 in C (K. 551, "Jupiter") (1843)', *Music Analysis in the Nineteenth Century*, i: *Fugue, form and style*, ed. I. Bent (Cambridge, 1993), 79–96

A. Lindmayr-Brandl: 'Mozarts frühe Tänze für Orchester', *MJb 1995*, 29–58

C. Eisen: 'The Salzburg Symphonies: a Biographical Interpretation', *Wolfgang Amadè Mozart: Essays on his Life and Music*, ed. S. Sadie (Oxford, 1996), 178–212

C. Eisen: 'Another Look at the "Corrupt Passage" in Mozart's G minor Symphony K550: its Sources, "Solution" and Implications for the Composition of the Final Trilogy', *EMc*, xxv (1997), 373–81

A. Kearns: 'The Orchestral Serenade in Eighteenth-Century Salzburg', *JMR*, xvi (1997), 163–97

E. Sisman: 'Genre, Gesture, and Meaning in Mozart's "Prague" Symphony', in Eisen, B 1997, 27–84

R. Münster: 'Zwei Versetten von Johann Ernst Eberlin im Gallimathias musicum KV 32', *MISM*, xlvi/3–4 (1998), 1–3

D. Chua: 'Haydn as Romantic: a Chemical Experiment with Instrumental Music', *Haydn Studies*, ed. D.W. Sutcliffe (Cambridge, 1999), 120–51

M: Concertos

C.M. Girdlestone: *Mozart et ses concertos pour piano* (Paris, 1939; Eng. trans., 1948, 3/1978)

G. Dazeley: 'The Original Text of Mozart's Clarinet Concerto', *MR*, ix (1948), 166–72; see also Kratochovíl, N 1956, and E. Hess, *MJb 1967*, 18–30

A. Hutchings: *A Companion to Mozart's Piano Concertos* (London, 1948, 2/1950/*R*)

H.C.R. Landon: 'The Concertos, II: their Musical Origin and Development', *The Mozart Companion*, ed. H.C.R. Landon and D. Mitchell (London and New York, 1956/*R*), 234–82

E.J. Simon: 'Sonata into Concerto: a Study of Mozart's First Seven Concertos', *AcM*, xxxi (1959), 170–85

H. Tischler: *A Structural Analysis of Mozart's Piano Concertos* (Brooklyn, NY, 1966)

I. Kecskeméti: 'Opernelemente in den Klavierkonzerten Mozarts', *MJb 1968–70*, 111–18

M.W. Cobin: 'Aspects of Stylistic Evolution in Two Mozart Concertos: K.271 and K.482', *MR*, xxxi (1970), 1–20

J. Kerman, ed.: *W.A. Mozart: Piano Concerto in C Major, K.503* (New York, 1970) [score and essays]

D. Forman: *Mozart's Concerto Form: the First Movements of the Piano Concertos* (London, 1971/*R*)

R. Strohm: 'Merkmale italienischer Versvertonung in Mozarts Klavierkonzerten', *Mozart und Italien: Rome 1974* [*AnMc*, no.18 (1978)], 219–36

C. Wolff: 'Zur Chronologie der Klavierkonzert-Kadenzen Mozarts', *Mozart und seine Umwelt: Salzburg 1976* [*MJb 1978–9*], 235–46

A.H. King: *Mozart String and Wind Concertos* (London, 1978/*R* with revs.)

S. McClary: 'A Musical Dialect from the Enlightenment: Mozart's Piano Concerto in G major, K.453, Movement 2', *Cultural Critique*, iv (1986), 129–69

A. Tyson: 'Mozart's Horn Concertos: New Datings and the Identification of Handwriting', *MJb 1987–8*, 121–37

R.D. Levin: *Who Wrote the Mozart Four-Wind Concertante?* (Stuyvesant, NY, 1988)

U. Konrad: 'Mozarts "Gruppenkonzerte" aus den letzten Salzburger Jahren: Probleme der Chronologie und Deutung', *Beiträge zur Geschichte des Konzerts: Festschrift Siegfried Kross*, ed. R. Emans and M. Wendt (Bonn, 1990), 141–57

W. Steinbeck: 'Zur Entstehung der Konzertsatzform in den Pasticcio-Konzerten Mozarts', ibid., 125–39

J. Kerman: 'Mozart's Piano Concertos and their Audience', *On Mozart: Washington DC 1991* (Washington DC, 1994), 151–68

M. Brück: *Die langsamen Sätze in Mozarts Klavierkonzerten: Untersuchungen zur Form und zum musikalischen Satz* (Munich, 1994)

M. Feldman: 'Staging the Virtuoso: Ritornello Procedure in Mozart, from Aria to Concerto', in Zaslaw, M 1996, 149–86

C. Lawson: *Mozart: Clarinet Concerto* (Cambridge, 1996)

J. Webster: 'Are Mozart's Concertos "Dramatic"? Concerto Ritornellos versus Aria Introductions in the 1780s', in Zaslaw, M 1996, 107–38
N. Zaslaw, ed.: *Mozart's Piano Concertos: Text, Context, Interpretation* (Ann Arbor, 1996)
S. Keefe: 'Koch's Commentary on the Late Eighteenth-Century Concerto: Dialogue, Drama and Solo/Orchestral Relations', *ML*, lxxix (1998), 368–85
S. Keefe: 'The Stylistic Significance of the First Movement of Mozart's Piano Concerto No.24 in C minor, K.491: a Dialogic Apotheosis', *JMR*, xviii (1999), 1–37

N: Chamber music

T.F. Dunhill: *Mozart's String Quartets* (London, 1927/*R*)
A. Einstein: 'Mozart's Ten Celebrated String Quartets', *MR*, iii (1942), 159–69
R.S. Tangemann: 'Mozart's Seventeen Epistle Sonatas', *MQ*, xxxii (1946), 588–601
W. Fischer: 'Mozarts Weg von der begleiteten Klaviersonate zur Kammermusik mit Klavier', *MJb 1956*, 16–34
J. Kratochvíl: 'Betrachtungen über die Urfassung des Konzerts für Klarinette und des Quintetts für Klarinette und Streicher von W.A. Mozart', *Leben und Werk W.A. Mozarts: Prague 1956* (Prague, ?1958), 262–7
S.T.M. Newman: 'Mozart's G minor Quintet (KV.516) and its Relationship to the G minor Symphony (KV.550)', *MR*, xvii (1956), 287–303
A.-E. Cherbuliez: 'Bemerkungen zu den "Haydn" Streichquartetten Mozarts und Haydns "Russischen" Streichquartetten', *MJb 1959*, 28–45
C. Bär: 'Die "Musique vom Robinig"', *MISM*, ix/3–4 (1960), 6–11
C. Bär: 'Die Lodronschen Nachtmusiken', *MISM* x/1–2 (1961), 19–22
E. Hess: 'Die "Varianten" im Finale des Streichquintettes KV593', *MJb 1960–61*, 68–77
K. Marguerre: 'Mozarts Klaviertrios', *MJb 1960–61*, 182–94
C. Bär: 'Zum "Nannerl-Septett" KV251', *Acta mozartiana*, ix (1962), 24–30
M. Whewell: 'Mozart's Bassethorn Trios', *MT*, ciii (1962), 19 only
A. Palm: 'Mozarts Streichquartett d-moll, KV421, in der Interpretation Momignys', *MJb 1962–3*, 256–79
C. Bär: 'Die "Andretterin-Musik": Betrachtungen zu KV205', *Acta mozartiana*, x (1963), 30–37
W. Kirkendale: 'More Slow Introductions by Mozart to Fugues of J.S. Bach?', *JAMS*, xvii (1964), 43–65
L. Finscher: 'Mozarts "Mailänder" Streichquartette', *Mf*, xix (1966), 270–83
A.H. King: *Mozart Chamber Music* (London, 1968)
K. Marguerre: 'Die beiden Sonaten-Reihen für Klavier und Geige', *MJb 1968–70*, 327–32
G. Croll and K. Birsak: 'Anton Stadlers "Bassettklarinette", und das "Stadler-Quintett" KV581: Versuch einer Anwendung', *ÖMz*, xxiv (1969), 3–11
W.J. Mitchell: 'Giuseppe Sarti and Mozart's Quartet K.421', *CMc*, no.9 (1969), 147–53
W.S. Newman: 'The Duo Texture of Mozart's K.526: an Essay in Classic Instrumental Style', *Essays in Musicology in Honor of Dragan Plamenac*, ed. G. Reese and R.J. Snow (Pittsburgh, 1969/*R*), 191–206
W. Hümmeke: *Versuch einer strukturwissenschaftlichen Darstellung der ersten und vierten Sätze der zehn letzten Streichquartette von W.A. Mozart* (Münster, 1970)

I. Hunkemöller: *W.A. Mozarts frühe Sonaten für Violine und Klavier* (Berne, 1970)
F. László: 'Untersuchungen zum Mozarts "zweiten" Opus 1, Nr.1', *MJb 1971–2*, 149–56
D.N. Leeson and D. Whitwell: 'Mozart's "Spurious" Wind Octets', *ML*, liii (1972), 377–99
R. Hellyer: 'Mozart's Harmoniemusik', *MR*, xxxiv (1973), 146–56
M. Flothuis: 'Die Bläserstücke KV439b', *Idomeneo Conference: Salzburg 1973* [*MJb 1973–4*], 202–10
J.A. Vertrees: 'Mozart's String Quartet K.465: the History of a Controversy', *CMc*, no.17 (1974), 96–114
D.N. Leeson and D. Whitwell: 'Concerning Mozart's Serenade in B♭ for Thirteen Instruments, K.361 (370a)', *MJb 1976–7*, 97–130
The String Quartets of Haydn, Mozart and Beethoven: Cambridge, MA, 1979 (Cambridge, 1980)
V.K. Agawu: *Playing with Signs: a Semiotic Interpretation of Classic Music* (Princeton, NJ, 1991) [on the string quintets]
W.-D. Seiffert: *Mozarts frühe Streichquartette* (Munich, 1992)
C. Eisen and W.-D. Seiffert, eds.: *Mozarts Streichquintette: Beiträge zum musikalischen Satz, zum Gattungskontext und zu Quellenfragen* (Stuttgart, 1994)
W. Pfann: '"Ein bescheidener Platz in der Sonatenform ...": zur formalen Gestaltung des Menuetts in den Haydn-Quartetten Mozarts', *AMw*, lii (1995), 316–36
W.-D. Seiffert: 'Mozart's "Haydn" Quartets: an Evaluation of the Autographs and First Edition, with Particular Attention to mm.125–42 of the Finale of K.387', in Eisen, B 1997, 175–200
J. Irving: *Mozart: the 'Haydn' Quartets* (Cambridge, 1998)

O: Keyboard music

F. Lorenz: *W.A. Mozart als Clavier-Componist* (Breslau, 1866)
H. Schenker: 'Mozart: Sonate a-moll', *Der Tonwille*, no.2 (1922), 7–24
H. Schenker: 'Mozart: Sonate C-dur', *Der Tonwille*, no.4 (1923), 19 only
N. Broder: 'Mozart and the "Clavier"', *MQ*, xxvii (1941), 422–32; repr. in Lang, B 1963, 76–85
H. Ferguson: 'Mozart's Duets for One Pianoforte', *PRMA*, lxxiii (1946–7), 35–44
W. Mason: 'Melodic Unity in Mozart's Piano Sonata K332', *MR*, xxii (1961), 28–33
K. von Fischer: 'Mozarts Klaviervariationen: zur Editions- und Aufführungspraxis des späten 18. und frühen 19. Jahrhunderts', *Hans Albrecht in memoriam*, ed. W. Brennecke and H. Haase (Kassel, 1962), 168–73
G. Croll: 'Zu Mozarts Larghetto und Allegro Es-dur für 2 Klaviere', *MJb 1964*, 28–37
H. Neumann and C. Schachter: 'The Two Versions of Mozart's Rondo K494', *Music Forum*, i (1967), 1–34
R. Rosenberg: *Die Klaviersonaten Mozarts: Gestalt- und Stilanalyse* (Hofheim, 1972)
W. Plath: 'Zur Datierung der Klaviersonaten KV 279–284', *Acta mozartiana*, xxi (1974), 26–30
Piano Quarterly, no.95 (1976) [Mozart issue]
M.H. Schmid: 'Klaviermusik in Salzburg um 1770', *Mozart und seine Umwelt: Salzburg 1976* [*MJb 1978–9*], 102–12

C. Wolff: 'Mozarts Präludien für Nannerl: zwei Rästel und ihre Lösung', *Festschrift Wolfgang Rehm*, ed. D. Berke and H. Heckmann (Kassel, 1989), 106–18

W.J. Allanbrook: 'Two Threads through the Labyrinth: Topic and Process in the First Movements of K.332 and K.333', *Convention in Eighteenth- and Nineteenth-Century Music: Essays in honor of Leonard G. Ratner*, ed. W. Allanbrook, J.M. Levy, and W.P. Mahrt (Stuyvesant, NY, 1992), 125–71

E.K. Wolf: 'The Rediscovered Autograph of Mozart's Fantasy and Sonata in C minor, K.475/457', *JM*, x (1992), 3–47

M.R. Mercado: *The Evolution of Mozart's Pianistic Style* (Carbondale, IL, 1992)

D.E. Freeman: 'Josef Mysliveček and Mozart's Piano Sonatas K.309 (284b) and 311 (284c)', *MJb 1995*, 95–109

D. Heartz: *Verona Portrait of Mozart and the Molto Allegro in G (KV72A)* (Ala, 1995)

J. Irving: *Mozart's Piano Sonatas: Contexts, Sources, Styles* (Cambridge, 1997)

C. Eisen: 'Mozart and the Sonata K.19d', *Haydn, Mozart, & Beethoven: Essays in Honour of Alan Tyson*, ed. S. Brandenburg (Oxford, 1998), 91–9

C. Eisen and C. Wintle: 'Mozart's C minor Fantasy, K475: An Editorial "Problem" and its Analytical and Critical Consequences', *JRMA*, cxxiv (1999), 26–52

P: Performing practice

R. Elvers: *Untersuchungen zu den Tempi in Mozarts Instrumentalmusik* (diss., Free U. of Berlin, 1952)

W. Fischer: 'Selbstzeugnisse Mozarts für die Aufführungsweise seiner Werke', *MJb 1955*, 7–16

H. Albrecht, ed.: *Die Bedeutung der Zeichen Keil, Strich und Punkt bei Mozart* (Kassel, 1957)

E. and P. Badura-Skoda: *Mozart-Interpretation* (Vienna, 1957; Eng. trans., 1962/*R* as *Interpreting Mozart on the Keyboard*)

P. Mies: 'Die Artikulationszeichen Strich und Punkt bei Wolfgang Amadeus Mozart', *Mf*, xi (1958), 428–55

A.B. Gottron: 'Wie spielte Mozart die Adagios seiner Klavierkonzerte?', *Mf*, xiii (1960), 334 only

C. Bär: 'Zum Begriff des "Basso" in Mozarts Serenaden', *MJb 1960–61*, 133–55

W. Gerstenberg: 'Authentische Tempi für Mozarts "Don Giovanni"?', ibid., 58–61

R. Münster: 'Authentische Tempi zu den sechs letzten Sinfonien W.A. Mozarts?', *MJb 1962–3*, 185–99

C. Bär: 'Zu einem Mozart'schen Andante-Tempo', *Acta mozartiana*, x (1963), 78–84

Z. Śliwiński: 'Ein Beitrag zum Thema: Ausführung der Vorschläge in W.A. Mozarts Klavierwerken', *MJb 1965–6*, 179–94

S. Babitz: 'Some Errors in Mozart Performance', *MJb 1967*, 62–89

C.-H. Mahling: 'Mozart und die Orchesterpraxis seiner Zeit', ibid., 229–43

E. Melkus: 'Über die Ausführung der Stricharten in Mozarts Werken', ibid., 244–65

H. Engel: 'Interpretation und Aufführungspraxis', *MJb 1968–70*, 7–16 [followed by proceedings of colloquium of Zentralinstitut für Mozartforschung, 1968, pp.20–46]

E. Melkus: 'Zur Auszierung der Da-capo-Arien in Mozarts Werken', ibid., 159–85

T. Harmon: 'The Performance of Mozart's Church Sonatas', *ML*, li (1970), 51–60

N. Zaslaw: 'Mozart's Tempo Conventions', *IMSCR XI: Copenhagen 1972* (Copenhagen, 1974), 720–33

M. Bilson: 'Some General Thoughts on Ornamentation in Mozart's Keyboard Works', *Piano Quarterly*, no.95 (1976), 26–8

F. Ferguson: 'The Classical Keyboard Concerto: some Thoughts on Authentic Performance', *EMc*, xii (1984), 437–45

J. Webster: 'The Scoring of Mozart's Chamber Music for Strings', *Music in the Classic Period: Essays in Honor of Barry S. Brook*, ed. A.W. Atlas (New York, 1985), 259–96

F. Neumann: *Ornamentation and Improvisation in Mozart* (Princeton, NJ, 1986)

J. Eppelsheim: 'Bassetthorn-Studien', *Studia organologica: Festschrift für John Henry van der Meer*, ed. F. Hellwig (Tutzing, 1987), 69–126

W. Malloch: 'Carl Czerny's Metronome Marks for Haydn and Mozart Symphonies', *EMc*, xvi (1988), 72–82

J.-P. Marty: *The Tempo Indications of Mozart* (New Haven, CT, 1988)

S. Rosenblum: *Performance Practices in Classic Piano Music* (Bloomington, IN, 1988)

H.M. Brown and S. Sadie, eds.: *Performance Practice: Music after 1600* (London and New York, 1989) [incl. articles by N. Zaslaw, M. Bilson, R. Stowell, D. Charlton, R.D. Levin, W. Crutchfield]

W. Crutchfield: 'The Prosodic Appoggiatura in the Music of Mozart and his Contemporaries', *JAMS*, xlii (1989), 229–74

P. Le Huray: *Authenticity in Performance: Eighteenth-Century Case Studies* (Cambridge, 1990)

R.L. Todd and P. Williams, eds.: *Perspectives on Mozart Performance* (Cambridge, 1991)

M. Bilson: 'Execution and Expression in the Sonata in E flat, K.282', *EMc*, xx (1992), 237–43

B. Brauchli: 'Christian Baumann's Square Pianos and Mozart', *GSJ*, xlv (1992), 29–49

D. Edge: 'Mozart's Viennese Orchestras', *EMc*, xx (1992), 64–88

C. Eisen: 'Mozart's Salzburg Orchestras', ibid., 89–103; repr. in *Journal of the Conductors' Guild*, xvii (1996), 9–21

W.-D. Seiffert: 'Anmerkungen zur Mozarts "Serenadenquartet"', *Gesellschaftsgebundene instrumentale Unterhaltungsmusik des 18. Jahrhunderts* (Tutzing, 1992), 105–18

C. Brown: 'Dots and Strokes in Late 18th- and 19th-Century Music', *EMc*, xxi (1993), 593–610

M.H. Schmid: 'Zur Mitwirkung des Solisten am Orchester-Tutti bei Mozarts Konzerten', *Basler Jb für historische Musikpraxis*, xvii (1993), 89–112

W.-D. Seiffert: 'Punkt und Strich bei Mozart', *Mozart als Text: Freiburg 1993*, 133–43

R. Maunder and D.E. Rowland: 'Mozart's Pedal Piano', *EMc*, xxiii (1995), 287–96

S. Rampe: *Mozarts Claviermusik: Klangwelt und Aufführungspraxis* (Kassel, 1995) [incl. discography and list of works]

R. Taruskin: *Text and Act: Essays on Music and Performance* (Oxford, 1995)

C. Eisen: 'The Orchestral Bass Part in Mozart's Salzburg Keyboard Concertos: the Evidence of the Authentic Copies', in Zaslaw, M 1996, 411–25

R. Riggs: 'Authenticity and Subjectivity in Mozart Performance: Türk on Character and Interpretation', *College Music Symposium*, xxxvi (1996), 33–58

Q: Reception

K. Werner-Jensen: *Studien zur 'Don Giovanni'-Rezeption im 19. Jahrhundert (1800–1850)* (Tutzing, 1980)

S. Dudley: 'Les premières versions françaises du *Mariage de Figaro* de Mozart', *RdM*, lxix (1983), 55–83

G. Gruber: *Mozart und die Nachwelt* (Salzburg, 1985, 2/1987; Eng. trans., 1991)

Mozart: origines et transformations d'un mythe: Clermont-Ferrand 1991 (Berne, 1994) [incl. J.-R. Mongrédien: 'La France à la découverte de Mozart ou le veritable enjeu d'une mythification (1791–1815)', 71–8; E. Kocevar: 'L'Évolution de la biographie mozartienne, de Friedrich von Schlichtegroll (1793) à Pierre Petit (1991): où la réalité devient mythe', 57–67; G. Gefen: 'Légendes maçonniques autour de Mozart', 43–55; J.-L. Jam: 'Wolfgang est grand, et Leopold est son prophète', 15–31]

P. Csobadi and others, eds.: *Das Phänomen Mozart im 20. Jahrhundert: Wirkung, Verarbeitung und Vermarktung in Literature, bildender Kunst und den Medien* (Salzburg, 1991)

B. Hiltner: *La clemenza di Tito von Wolfgang Amadé Mozart im Spiegel der musikalischen Fachpresse zwischen 1800 und 1850* (Frankfurt, 1994)

H. Jung, ed.: *Mozart: Aspekte des 19. Jahrhunderts* (Mannheim, 1995)

F. Senici: *La clemenza di Tito di Mozart: i primi trent'anni (1791–1821)* (Turnhout, 1997)

Index